Aircraft Systems

Understanding Your Airplane

**TAB
PRACTICAL
FLYING SERIES**

No. 2423
$27.95

Aircraft Systems

Understanding Your Airplane

David A. Lombardo

TAB BOOKS Inc.
Blue Ridge Summit, PA

Notice

The information contained herein was originally published as a series in *Private Pilot* magazine.

FIRST EDITION
FIRST PRINTING

Library of Congress Cataloging-in-Publication Data

Lombardo, David A.
   Aircraft systems : understanding your airplane / by David A. Lombardo.
      p.   cm.
  Includes index.
  ISBN 0-8306-1823-6   ISBN 0-8306-0823-0
  1. Airplanes. 2. Private flying. I. Title.
TL670.L66 1988   88-17069
629.134—dc 19   CIP

TAB BOOKS Inc. offers software for sale. For information and a catalog, please contact TAB Software Department, Blue Ridge Summit, PA 17294-0850.

Questions regarding the content of this book should be addressed to:

    Reader Inquiry Branch
    TAB BOOKS Inc.
    Blue Ridge Summit, PA 17294-0214

Cover photograph courtesy of Cessna Aircraft Company

To Dad

Contents

PART 2: ELECTRICAL SYSTEM

Acknowledgments

The path toward this book begins with Kevin Murphy, self-proclaimed world's greatest pilot and aviation author. Kevin badgered me for years to start free-lance writing for aviation publications. He and I had taught together for some time and felt there was an unmet need for technically oriented articles within the general aviation community. In 1980, while teaching a flight instructor refresher clinic in St. Louis, I had the good fortune to meet and become friends with the noted aviation author John Lowery. John convinced me that not only was there the need, but that I should help meet that need. Over the years, John has been both an inspiration and a sounding board.

Every magazine I contacted expressed an interest in technical material; they were apparently inundated with operational, flight safety, and airplane review articles. What the editors of aviation magazines appeared to have trouble finding were articles about aircraft systems. How do they work? How should they be preflighted? How does a pilot troubleshoot a system inflight? What preventive maintenance is necessary? As a professional pilot with an Airframe and Powerplant Mechanic's certificate, I might not have appreciated that need. It wasn't until I did some flight instructing outside of a university environment that I began to realize how little aircraft systems information was available to the average general aviation pilot.

While avionics equipment typically comes with a detailed manual, aircraft systems do not. Every modern aircraft must have a Pilot's Operating Handbook (POH), but what the POH doesn't tell you can literally fill volumes. Then, too, there are optional and retrofit systems that are not discussed in the POH at all. After years of reading about aircraft accidents and conducting safety seminars, I have become impressed with one important fact: Most accidents are the result of ignorance, not stupidity, and ignorance can be fought with information and interest.

I wrote and submitted an article to *Private Pilot* magazine and almost immediately received a reply from Editor Dennis Shattuck. He wanted that article and was interested in more of the same. If there is such a thing as a writer's guardian angel, Dennis is it. He teaches, cajoles, and offers both solicited and unsolicited advice with one eye on good copy and the other on preserving the writer's often brittle ego. There is no question in my mind that whatever success I have had in the field of writing is in no small part attributed to Dennis. The aircraft systems series ran for 2½ years and I believe was successful primarily because it served as the Rosetta Stone between the manufacturer and the pilot, translating what I call "technobabble" into English.

Free-lance technical writing is an incredibly time-consuming effort. I began while a professor in the Department of Professional Aviation at Louisiana Tech University and found the faculty and students to be highly supportive of my efforts. Al Miller, the Department Head, was not only instrumental in assuring me I had the opportunity to write but was the finest employer I have ever known.

The idea of publishing the articles as a book came from Pat Brown, a lecturer at Texas State Technical Institute. Pat told me he was using my articles as part of the teaching materials for TSTI's aircraft systems course and encouraged me to get them all into a book. For that I am indebted to Pat.

Those to whom I have turned repeatedly for technical expertise over the years include Wilson Jones and John Filhiol of Louisiana Tech University and Weldon Garrelts of the University of Illinois Institute of Aviation. Bill Geibel, also at the University of Illinois, has been very supportive and an excellent maintenance resource person. And, of course, I thank all the industry personnel who have provided me with technical information, explanations, and illustrations. Illustrations and photos, not otherwise credited, are provided through the generosity of Norman Ridker and *Private Pilot* Magazine.

Vicki Cohen, who has for some time put up with my quirks and idiosyncrasies, has been the one person that helped me maintain both my sense of humor and perspective while trying to cope with life's foibles, a demanding job, and a burning desire to get this book to press. I thank her for just being there.

And finally, I thank my Dad, to whom this book is dedicated. Through good times and bad, he has always encouraged and supported my every endeavor. No child could have asked for more love, no adult for more approval. Through 10 years of college and countless moves, his timid question was only, "where is this all going?" And finally my answer is, "here."

Introduction

Since June 17, 1969, I have been continuously involved either directly, or indirectly, in aviation/aerospace education. On that day, I received my first dual flight in a Cessna 150, and though both the airplane and the airport are gone, the memory is as fresh as if it happened yesterday. Flying is something that gets in a person's blood; even an extended absence cannot prevent a person from saying "I'm a pilot." But the love of flying should never be confused with the technique of flying.

Thousands of pilots have safe flights every day, but that doesn't mean they fly safely everyday. Many pilots assume that because they have both a large and an equal number of takeoffs and landings, they are safe pilots. Unfortunately that is not the case, and NTSB accident statistics prove it. Safety is a relative concept; it is directly proportional to skill, knowledge, and judgment.

This book stresses knowledge of general aviation aircraft systems. It isn't as if systems have been totally avoided by authors; there are some excellent texts for use in aircraft maintenance schools. Many universities and other professional pilot programs have been using these texts, but unfortunately there are several serious drawbacks. Such texts always go into far greater detail than is warranted, or even desired, for pilot education. The material often requires specialized knowledge of physics, mathematics, and chemistry not covered in the book. The necessary material is seldom available in a single text, typically requiring two or three. And in virtually every case, the price of such texts is high. I decided that a reasonably priced, pilot-oriented book was necessary.

Aircraft Systems—Understanding Your Airplane does not require the reader to have any special knowledge. When necessary, terms are defined and concepts are clarified.

The book is as useful to the student pilot as it is to the experienced one. Because each chapter covers a different aircraft system, it can be read from cover to cover or used as a reference book.

For quite a few years, I have been a very active aviation safety public speaker and volunteer FAA Accident Prevention Counselor. Recently, I have been faced with the awful truth that despite the tremendous effort made by the FAA and other aviation safety-oriented groups, it's not possible to reach all the pilots through safety seminars. It is my deepest hope that this book and others like it will be read by every general aviation pilot. Knowledge is the greatest tool a pilot can ever possess.

David A. Lombardo
Champaign, IL
May, 1988

1

Know It's Airworthy Before You Fly It!

Someone once said, "Don't meet trouble halfway, it is quite capable of making the entire journey." Most pilots unknowingly court disaster by doing only a basic preflight walk-around inspection each time they fly. A thorough preflight is the foundation of a safe flight, and the checklist should always be consulted during or after the walkaround to assure that nothing has been missed. But there is more to consider: *aircraft security* and *attrition*, the two primary reasons for the airframe airworthiness preflight.

Aircraft security implies everything from making sure the propeller is where it belongs to ensuring that the flightline attendant hasn't driven the fuel truck into the wing. It includes checking to see that no one has let the air out of the tires and that there's enough fuel and oil to make the trip. It's what we've come to think of as the preflight inspection. Attrition, on the other hand, refers to the general condition of the aircraft.

FIVE AGING ELEMENTS

Five elements constantly age an airplane: weather, friction, overloads, heat, and vibration. Every preflight should be conducted with the effects of these five elements in mind.

Weather

Weather has a whole arsenal of weapons. Temperature, humidity, rain, wind, snow, and ice constantly erode a shiny, new airplane. The more obvious weather-related pre-

1

flight items, such as 2 inches of snow on the wings or the effects of a hot day on takeoff performance present no problem to the diligent pilot. But it's the subtle weather effects—the long-term ones—that tend to take the biggest toll. When comparing costs for storing aircraft, tiedowns certainly seem more practical than hangaring. But over a period of time, each day's sun, wind, and dust chisel away at the paint and windows.

Friction

Friction between moving parts, such as the hinges on ailerons, elevators, and rudder cause continuous wear that is seldom checked on a preflight. Simply moving a control surface does not adequately check for wear; it assures freedom of travel that might otherwise be restricted by gust locks, broken cables, or other problems preventing full deflection. Control surface movement should be evaluated with respect to the type and quality of movement; surfaces should feel reasonably solid and not sloppy. Less obvious areas of friction occur in such places as the engine, where cooling baffles rub the cowling, slowly eating their way through the aluminum.

Overloads

Overloads can cause deformation and/or failure of the structure. Physically overloading the aircraft with people, fuel, and other goodies, then exceeding the maximum load limit with a steep bank turn, sharp nose over, or other high-G-producing maneuver causes undue stress. The airframe can also experience overload conditions significantly below maximum gross weight during a hard landing, in turbulence while above maneuvering speed, or by a close encounter with a thunderstorm. Understandably, after-the-fact detection is usually difficult.

Heat

Heat comes in two forms: *direct* and *indirect*. Heat from the exhaust system is considered direct. The danger is the potential for leaking carbon monoxide into the aircraft's heating system and, therefore, into the cabin. Exhaust welds should be periodically checked and an inexpensive carbon monoxide detector should be put into every airplane so the level of carbon monoxide can be checked regularly. Detectors should be changed frequently because they gradually lose their effectiveness. Indirect heat comes from inadequate cooling of an engine. It can be detected through preflight and inflight symptoms, such as high oil and cylinder head temps, an odor of burned oil or hot rubber during engine operation or shortly after shutdown, and the auto-ignition of an engine after shutdown. Any such symptom indicates a potentially severe engine problem and requires immediate consultation with a mechanic. The blistering of cowling paint over the engine during starting is a sure sign of an induction fire. If this occurs, continue cranking the engine with the starter to draw the fire back into the cylinders. Shutting down the engine during an induction fire allows the fire to burn inside the cowling, which could result in irreparable damage.

Vibration

Although there are many normal vibrations in an aircraft, unusual vibrations signal danger. They can be caused by ice that disturbs airflow over a surface or antenna, a loose control surface, airspeed in excess of designed normal operating speed, or an engine- or propeller-related malfunction. Seldom can a vibration-causing problem be cured in flight. Generally the best you can hope for is to subdue it by reducing power and/or airspeed. Be careful to avoid any increase in load factor and land as soon as it's practical.

EXTERNAL PREFLIGHT

On the ground, on the surface, and under the surface are the three areas for consideration during the exterior preflight inspection. When approaching the aircraft, look at it from a distance and ask yourself if anything looks unusual. I remember watching a student preflight an Aeronca Champ that clearly had one wing recently modified—by GMC—into a swept-back position; a fuel truck had backed into it. From up close the student really could not tell that the two wings were at slightly different angles. Moving closer to the aircraft, observe the ground under and around it. Are any loose parts lying there, such as bolts or screws? What about a pool of liquid under the fuselage that could indicate a fuel, oil, or hydraulic leak? Sometimes such leaks leave tell-tale traces on the underside of the airplane, but not always.

Cracks and Wrinkles

There are numerous special considerations when inspecting the exterior airframe. A crack in the structural part of the skin is reason to call an A&P mechanic. Cracks in non-structural parts such as fairings and wheel pants are not necessarily grounding items, provided the airflow will not cause the crack to open farther, possibly causing a break in the piece and resulting in severe airflow disruption. A flapping piece of fairing can be alarming and can cause aerodynamic problems. A neat, single layer of duct tape may be used to temporarily secure small cracks in non-structural areas. But it's always best to have a mechanic look it over before flight.

Wrinkles in aluminum skin are often the result of an overload, such as excessive inflight G loads. Once stretched, the skin will flex slightly when pressed, like pushing on the side of a gas or oil can, hence the name "oil canning." Abrupt maneuvering, turbulence, or improperly executed aerobatic maneuvers are typical culprits. Low-wing aircraft with landing gear attached to the spar are particularly susceptible to hard landings, because the overload is transferred from the landing gear to the spar, which bends slightly under the load, and causes the aluminum skin to stretch. This is one of the first things to check when considering the purchase of a used low-wing trainer. These aircraft experience a higher-than-average incidence of hard landings, or as one student pilot put it, ''any landing you walk away from is a good landing.'' Student pilots notwithstanding, some ''give'' in the surface is to be expected, but considerable flex could indicate severe damage to the wing and spar. Another indication of probable damage is missing or popped rivets. A popped rivet is easy to detect by the black oxide which will seep out from under it. Simply sight along a line of rivets and look for small, black stains. Press on the surface next to the rivet to confirm it is loose. Again, an A&P mechanic should be consulted because popped or missing rivets can indicate a potentially deeper, more serious problem.

Windows

Contemporary aircraft windows are made of Plexiglas, and when maintained properly transmit 90 percent of available visible light, which is better than glass. Unfortunately the poor hardness qualities of Plexiglas make it very susceptible to scratching. Along with distracting reflections, scratches can cause severe glare problems as they catch the sun at varying angles. Proper care of windows can go a long way toward preventing such problems. You should never clean a window with any coarse cloth or paper towel. Be particularly watchful for the ''efficient'' flightline attendant who rushes out with a bottle of chemical cleaner and a dirty rag to clean your windshield. Chemical cleaners, particularly those used in the home, can react with and damage Plexiglas. For the most part, it is best to refuse the offer to have your windows cleaned and do it correctly yourself. The proper technique is to first soak the deposits and dirt on the window with a very mild soap and water to soften them. Then after taking off rings, watches and other hand and arm jewelry, gently wipe off the wet window using a very soft, clean, absorbent cloth.

You should consider replacing windows when there is a distinct discoloration or milky appearance. Known as *crazing*, the window is extensively covered by tiny little separations or hairline cracks that significantly weaken its strength. Any window with a visible crack should be immediately referred to a mechanic. If caught early enough, it's possible to stop-drill a crack to prevent further damage. Some aircraft owners, in an attempt to save wear and tear on windows, install external window covers, but their value is debatable. If used in a hangar only, they might keep the birds from using the window for target practice; but outside, the cover can flap wildly during high winds, beating dust against the window and accelerating its deterioration. Those who are concerned about cabin heat might choose to use internal Velcro-mounted covers that serve as both heat shield and a security device, preventing curious onlookers from sizing up your avionics investment. For more about window care, see Chapter 26, Clear Facts about Transparencies.

Cleanliness

Besides the aesthetic satisfaction that results from having a clean airplane, there are several very practical reasons. A clean airplane shows tell-tale traces of fluid leaks, popped rivets, and other problems more readily than a dirty one. Often a leak can be traced along the fuselage to its source. From an economical point of view, the washed and waxed airplane is aerodynamically cleaner and you can expect to increase your cruise speed by a couple of knots.

While on the subject of increasing speed, let's take a look at those speed fairings the manufacturers put over non-retractable landing gear. You know, the ones your mechanic curses when it's time to change a tire? The folks up North call them ''snow catchers'' and usually remove them permanently after the first snowfall. Not only do they increase cruise airspeed, but many cruise performance charts such as for the Cessna 152 are calculated with the fairings on. Read the fine print closely to see if the computations account for the fairings.

Vents and Openings

At the risk of sounding like a fisherman talking about the one that got away, a friend of mine ran out of fuel inflight (through no fault of his own). The interruption was due to a fuel tank air vent that was blocked by a mud dauber's nest. The old axiom ''what you don't know can't hurt you'' seldom holds true in aviation.

Vents and openings—especially the pitot tube and static port—should be checked for foreign objects that might be lodged inside. A former flight instructor used to have a pre-flight pet trick. He would put a toothpick in the pitot tube of his Piper Cherokee to see if his students actually got down on all fours and checked it. One day, that student's lesson was canceled at the last minute, and another student took the airplane up solo, toothpick and all; it was quite a ride for the novice. There are far too many stories about blockages caused by dirt, ice, snow, and other foreign objects to bypass a thorough look.

Life Expectancy

Many pilots think of the life expectancy of their airplane in two ways; the airframe is forever and the powerplant until TBO (time between overhaul). Inherent in that belief is the assumption that the accessories are either indestructible or at worst are certainly good for the life of the engine. Nothing could be further from the truth. Certainly, the closer an engine gets to its TBO, the greater the scrutiny it should receive during preflight—and during flight. Most engines have TBO's of between 1,200 and 2,000 hours; other aircraft parts have significantly shorter life spans. For instance, most propellers are rated at less than engine TBO (1,500 hours is common). Fuel boost pumps range from 700 to 1,500 hours. Replacing the filter every 100 hours (something few pilots ever do), helps a dry vacuum pump to run from 400 to 1,000 hours (the wet type typically has a longer life). Mufflers frequently do not go beyond 1,000 hours, and alternators, which have a nasty habit of going out while flying in clouds, generally give up the ghost around 1,200 hours. Component life expectancy varies between manufacturers. And, the way you operate your aircraft can add to or detract from the manufacturers recommended time between overhaul or replacement. The trick is to have a fair idea how long these components should last and begin to cast a suspicious eye upon them from preflight through shutdown.

Rust and Corrosion

Rust is a form of oxidation caused by the reaction of ferrous metals such as iron and steel with oxygen. When it forms in an area that can be readily seen, it is easily detectable by its typically reddish discoloration. In its early stage, rust can be cleaned off with a little elbow grease and a rag. Or, if necessary, a very mild abrasive such as emery cloth can be used and the surface painted to prevent future rusting. A reddish-brown crustiness indicates advanced rust, and removing it will reveal pitting in the aircraft's surface. Once pitting occurs, a qualified mechanic should be consulted to evaluate the situation. A more serious situation occurs when the rusting takes place in an area not easily observable by the pilot, allowing it to develop to its advanced stage. This commonly happens in the belly of the aircraft, inside the control surfaces, on the wings and empennage, and inside steel-tube members of floatplanes. Aircraft operating routinely in wet environments and all seaplanes should be periodically inspected for rust by a qualified mechanic.

Corrosion, also a form of oxidation, is an electrochemical process involving nonferrous materials, such as aluminum, copper, and magnesium. The presence of some chemicals such as battery acid, insecticides, and fertilizers accelerate corrosion where dissimilar metals exist side-by-side. It is usually observed as a grayish-white powder. As with rust, if it can be removed completely with a gentle cleaning, you've probably caught it in time. But the presence of pitting indicates possible permanent damage and a mechanic must be consulted. Corrosion under paint or other surfaces can show in surface flaking, pitting, blistering, or bubbles. Areas that should be inspected regularly include: engine exhaust stacks and exhaust areas; battery compartments and battery vents; landing gear and wheel-well areas; surface skin seams and piano hinges on control surfaces and access doors.

INTERIOR PREFLIGHT

Before conducting the interior preflight, take a close look around the cabin. It should be neat, secure and clean. You should be able to find what you need with a minimum diversion of attention during flight. Items not securely fastened can become potentially lethal projectiles. And, finally, if a lot of dust is floating in the cabin during flight, it can cause eye and respiratory irritation. As a student pilot, I found the airplanes were so dirty at my local FBO that for a long time I logged two sneezes for every touch-and-go.

While looking around inside, check the seats, windows, and door latches. There have been several fatal accidents as a result of the pilot seat sliding all the way back just after takeoff. Another common problem, though not so crucial, is the window popping open in flight. This one tends to frighten passengers more than anything, but a door popping open in flight could cause problems. Most aircraft are capable of flying with an open door, however it could be a problem in some. Another potentially bad situation can be caused by baggage and access doors opening in flight. In the single-engine airplane, a baggage door might not present too great a hazard. A nose cowling, however, might make an attempt at getting into the front seat with you. In the twin, forward baggage doors can present a very serious threat. The prospect of a door opening and allowing baggage to spill into a prop is frightening, not to mention the possibility that the door itself could break loose and fly into the prop.

PREVENTIVE MAINTENANCE

Routine preventive maintenance keeps most problems in check. It is defined in FAR (Federal Aviation Regulation) Part 1 as "simple or minor preservation operations and the replacement of small standard parts not involving complex assembly operations." It is corrective action taken before it becomes necessary to make more complex repairs. Preventive maintenance, as outlined in Federal Aviation Regulation 43 may be approved for return to service by "a person holding at least a private pilot certificate." See TABLE 1-1 for those repairs viewed as preventive maintenance. Most of these items are easily accomplished by anyone with a basic mechanical aptitude, but a good rule of thumb is "if you aren't sure what you are doing, don't do it."

WHAT TO TELL THE A&P

At some point, no matter how good the preventive maintenance, no matter how much you baby your airplane, you are going to need the services of your local, friendly airframe and powerplant mechanic. If there is one thing a mechanic cringes at, it's a pilot who says "plane don't work right, fix it." One hapless pilot said that and returned to a bill of $2,000 which did not include fixing the problem he was originally concerned about. It is important to accurately define the problem and its location for the mechanic to provide quick, efficient, and reasonably priced work. The statement of the problem should reflect exactly what symptoms you observe and under what conditions you observed them. Telling your A&P that the airplane has a vibration is like telling your doctor you're sick. You must be specific: a low frequency vibration coming from under the airplane in icing conditions points to a very different problem than a high frequency vibration in a cruise descent coming from the area of the engine cowling. Pertinent information should include the following:

1. A precise description of the problem
2. The severity of the problem
3. The power setting at time of problem

Table 1-1. Extract from FAR Part 43, Appendix A (Major Alterations, Major Repairs, and Preventive Maintenance)

(C) *Preventive maintenance.*

Preventive maintenance is limited to the following work, provided it does not involve complex assembly operations.

1. Removal, installation, and repair of landing gear tires.
2. Replacing elastic shock absorber cords on landing gear.
3. Servicing landing gear shock struts by adding oil, air, or both.
4. Servicing landing gear wheel bearings, such as cleaning and greasing.
5. Replacing defective safety wiring or cotter keys.
6. Lubrication not requiring disassembly other than removal of non-structural items such as cover plates, cowlings, and fairings.
7. Making simple fabric patches not requiring rib stitching or the removal of structural parts or control surfaces.
8. Replenishing hydraulic fluid in the hydraulic reservoir.
9. Refinishing decorative coating of fuselage, wings, tail group surfaces (excluding balanced control surfaces), fairings, cowling, landing gear, cabin, or cockpit interior when removal or disassembly of any primary structure or operating system is not required.
10. Applying preservative or protective material to components where no disassembly of any primary structure or operating systems is involved and where such coating is not prohibited or is not contrary to good practices.
11. Repairing upholstery and decorative furnishings of the cabin or cockpit interior when the repairing does not require disassembly of any primary structure or operating system or interfere with an operating system or affect primary structure of the aircraft.
12. Making small, simple repairs to fairing, nonstructural cover plates, cowlings, and small patches and reinforcements not changing the contour so as to interfere with proper airflow.
13. Replacing side windows where that work does not interfere with the structure or any operating system such as controls, electrical equipment, etc.
14. Replacing safety belts.
15. Replacing seats or seat parts with replacement parts approved for the aircraft, not involving disassembly of any primary structure or operating system.
16. Troubleshooting and repairing broken circuits in landing light wiring circuits.
17. Replacing bulbs, reflectors, and lenses of position and landing lights.
18. Replacing wheels and skis where no weight and balance computation is involved.
19. Replacing any cowling not requiring removal of the propeller or disconnection of flight controls.
20. Replacing or cleaning spark plugs and setting of spark plug gap clearance.
21. Replacing any hose connection except hydraulic connections.
22. Replacing prefabricated fuel lines.
23. Cleaning fuel and oil strainers.
24. Replacing batteries and checking fluid level and specific gravity.
25. Removing and installing glider wings and tail surfaces that are specifically designed for quick removal and installation and when such removal and installation can be accomplished by the pilot.

4. The engine instrument indications during problem
5. The corrective measures you attempted and the results
6. The flight condition at the time of the problem (cruise, slow flight, descent, etc).
7. The approximate gross weight and loading
8. The outside air temperature
9. The presence of visible precipitation (rain, ice, snow, etc.)
10. The type and severity of any damage
11. The known causes of damage

Another pet peeve of many mechanics is the pilot who says, "There's a rivet missing. See ya after lunch." In describing problems, be specific about locations and use correct terminology, such as forward or aft, port or starboard (left and right are acceptable, as viewed from the pilot's seat), upper or lower, inboard or outboard, and leading edge or trailing edge. For example, "there is a 1-inch crack on the lower outboard trailing edge of the port (left) aileron."

An airworthy flight results from a thorough preflight. Check for signs of security and attrition, practice preventive maintenance, be aware of airframe and engine performance during flight, and when necessary, get proper maintenance by a qualified mechanic.

2

Parts Is Parts?
Not So
with Airplanes!

There are two categories of guidelines for determining the airworthiness of an aircraft: Type Certificate Data Sheets (TCDS) and Supplemental Type Certificates (STC).

At the origin, an aircraft product is certified by the FAA under the specification listed in the TCDS. If it is maintained properly, it is considered airworthy. The original specification may be modified later by an Airworthiness Directive (AD). Sometimes, after a new product has been on the market for a while, design flaws or operational problems are discovered. An AD is then issued requiring the owner to correct the problem as outlined.

ADs have a compliance date, after which the aircraft no longer is considered airworthy. Unless there is a very serious problem, the time allowed to make the fix is usually more than adequate. Few owners are happy to receive an AD note in the mail, but aside from some inconvenience and expense to the owner, the system works well.

The second guideline is the STC, which is issued by the FAA when the product (aircraft) has been altered from the original TCDS. All changes must be approved by the FAA in an STC. Once the changes are approved, the aircraft or product is considered airworthy and may fall entirely under an STC. For instance, an STC is required if the owner of a single-engine aircraft wants to install a backup vacuum system that was not originally approved by the airframe manufacturer. Often, the manufacturer of the vacuum system modification already has the STC, which specifies the make and model of the aircraft for which it is approved.

Many pilots believe compliance with required periodic maintenance and adherence to TCDS, ADs, and STCs is all that is required to keep an aircraft airworthy, but such compliance might not be enough. Parts used for compliance must also be approved by the FAA.

How many pilots would ground their aircraft and make an appointment with the local shop when they discover that a small screw is missing from a wingtip fairing or a bolt is missing from the seat track? More likely than not, the pilot goes to his home workshop and digs into his coffee-can collection of old screws to find a near duplicate. Such low-cost parts can be very attractive, especially when doing your own work, and would seem to present no problem. In reality, use of unapproved parts automatically invalidates the aircraft's airworthiness certificate.

Fortunately, very few maintenance facilities would ever consider using unapproved parts, but the temptation to use "approved" low-cost parts is tremendous.

THREE TYPES OF AIRCRAFT PARTS

There are three types of parts for aircraft: Original Equipment Manufacture (OEM), Parts Manufacturer Approval (PMA), and *bogus*.

The screw a pilot fishes out of the workshop coffee can could be of the bogus variety. Bogus parts are not limited to such obvious origins. Some unscrupulous operators dig into their own coffee cans and try to sell you the parts as FAA-approved. A few even will go so far as to scribe official looking numbers on the part and/or alter it to resemble approved parts, despite potentially stiff fines resulting from such violation of the FARs. Fortunately, this is as rare as it is illegal.

More bogus parts find their way into civil aircraft through the salvage and military surplus markets. The uninformed owner assumes that if it came off another aircraft, especially a military one, it must be acceptable. But there is no guarantee that the aircraft it was salvaged from didn't overstress or otherwise damage the part. There also is a question of storage, exposure to a harmful environment, and even the compatibility of the original manufacturing process with its intended use.

There are only two types of FAA-approved parts: OEM and PMA. Original Equipment Manufacturer (OEM) parts are those designed and constructed for the manufacturer of the original piece of equipment. These parts are issued a TCDS. For instance, when you purchase a brand-new Lycoming engine from its manufacturer, the entire engine consists of OEM parts. Some day a mechanic will discover a part needs to be replaced in that engine. When that happens, there are two legal choices the mechanic can make: install an OEM or a PMA part. (Any subsequent manufacturer who desires to build and offer a replacement part or product must get a PMA from the FAA before that part is legal.)

According to FAR Part 45.15, PMA parts must be marked with the letters FAA-PMA, name, trademark or symbol, part number, and name and model designation on which the product is eligible for installation. Parts too small to be marked with an etching, such as a ball bearing, must have a tag (even if it is only enclosed in a plastic bag with the part) or a similar method of identification. These methods of marking make it difficult for the unscrupulous dealer to sell bogus parts to the aware aircraft owner.

To be certain of acceptability when purchasing a part, you could ask for evidence of a PMA letter of approval, listed by part number and eligibility for the type of product on which it may be used. Every FAA-PMA part has one on file. Unfortunately, most distributors are unlikely to have such a letter as it is issued to the manufacturer. Distributors might want to purchase a copy of the "Parts Manufacturer Approvals" publication from the U.S. Superintendent of Documents. Concerned aircraft owners could do the same, but the most practical course of action is simply to deal with an established and reputable dealer.

Why should PMA parts be considered? Some manufacturers think they have come up with a better idea than the original manufacturer. They offer what they consider to be an improved part. To get a PMA, they must convince the FAA that the part is as good as the original, and they must also obtain an STC incorporating the changes or improvements.

There is quite a controversy raging between OEMs and PMA manufacturers. The problem isn't so much who makes the part; airframe and engine manufacturers don't manufacture all of their own parts. It is not uncommon for an independent contractor to build a part under contract with the original manufacturer, using the OEM blueprints and specifications. For instance, the pistons and rings that come with a new Continental engine actually are built to Continental's specifications by an independent contractor. The problem is that OEMs dislike replacement parts being built by another manufacturer as PMAs.

OEMs sometimes think PMA manufacturers typically do little or no research and testing of their own prior to getting FAA approval. Some manufacturers claim FAA enforcement of PMA quality control practices is not as strict as that for OEMs. But virtually all OEMs express concern about liability when their product is overhauled with PMA parts. It's a fact of life that when there's trouble with a component, the OEM gets sued, regardless of whose parts went in it during its last overhaul.

Some manufacturers would have you believe PMA parts are not as good as OEM parts. COMSIS Corp. of Wheaton, Maryland, did an FAA-sponsored evaluation of the PMA process. "There is no significant problem with PMA parts," the study reported. But the nagging problem of liability continues to crop up and cause significant concern among OEMs. In 1985, Teledyne Continental Motors issued TBO (time between overhaul) Service Bulletin #85-13, informing owners that the Continental engine TBO will apply only to engines that have used OEM parts from the beginning. Such a move by Teledyne could significantly affect the small overhaul shops who use PMA parts to keep overhead down. Larry Johnson, Teledyne Continental's director of marketing, summed up Teledyne's opinion when he said, "We just don't know how that engine will work with parts we have no control over."

Of course there are two sides to every dispute. PMA dealer Chuck Dedmon, President of Superior Air Parts in Addison, Texas, has a slightly different view. Superior, which has the world's largest inventory of FAA-PMA parts for Continental and Lycoming piston engines, has its own quality-control department and laboratory. About 80 percent of what it sells is built to the manufacturers' specifications; the other 20 percent of its business is the distribution of other manufacturer's parts. According to Dedmon, everything sold as a Superior part must go through Superior's quality control department.

Some companies, such as Eaton which makes hydraulic lifters, have their own quality control departments. But as a matter of routine, Superior still puts the outside vendor parts through its own in-house quality control process. Dedmon says the whole controversy about PMA parts stems from the fact that prior to 1967, Continental and Lycoming had no competition. Then, Superior entered the market and began selling approved parts at a lower cost.

The question of whether a PMA part is as good as the original is difficult to resolve. A review of FAR Part 43.13b hardly clears up the matter. According to this regulation, "Each person maintaining, altering or performing preventive maintenance shall do that work in such a manner and use materials of such a quality that the condition of the aircraft, aircraft engine, propeller, or appliance worked on will be at least equal to its original or properly altered condition . . ."

PART MANUFACTURING PROCESSES

When asked if a PMA part is as good as an OEM part, Walter Horn, manager of the FAA's Chicago Aircraft Certification Office, quickly responded: "Does the part have FAA-PMA on it? If so, it's as good as original." No indecision there. He spent some time describing the two-step process required to get PMA approval. First, there is *design approval*, also known as *data approval*, by the FAA Engineering Aircraft Certification Office (ACO). This can be accomplished by any one of three methods: identicality, licensing agreement with the OEM, or by designing and building your own substitute part.

Identicality Method

In the identicality method, the manufacturer requesting PMA approval must prove to the FAA that the part is identical in every way to the OEMs. This is done by submitting drawings and specifications to the FAA for its review. According to many OEMs, this is where the problem comes in. They argue that these specifications are obtained through reverse engineering; the prospective manufacturer buys the original part, studies it, measures it and develops a set of drawings and specifications. The expense of initial research and development being paid by the OEM, the new manufacturer now can sell the part for less.

Reverse engineering does take place, but when the FAA says the drawing and specifications must match the OEMs, they mean match. Any deviation at all is cause for immediate rejection, and because the specification is being compared to proprietary information, the FAA cannot indicate what area or areas are unacceptable. Many things in the manufacturing process, such as hardening processes and heat treatments, are not readily discernible simply by looking at the part. Acceptance by identicality means the part will be an exact duplicate in every way.

Licensing Approval

The second method of getting design approval—a licensing agreement—is very straight-forward, with drawing and specifications supplied by the OEM. This practice is common, especially for airframe manufacturers who may have a difficult time keeping up with normal production schedules, let alone spare parts.

Original Design and Approval

The third method of getting design approval is for the new manufacturer to design and build the substitute part from scratch. The manufacturer would have to run its own tests, subjecting the part to loads, pressure, and environmental stresses to substantiate to the FAA that the new part is of equivalent strength, durability, and safety. This method also implies the part is as good as the original.

After acceptance by one of the three methods, final approval must be made by one of the FAA's Manufacturing Inspection District Offices (MIDO). These are not engineers, rather they are individuals who know the manufacturing processes and techniques necessary to produce the part. Their task is to evaluate the company's ability to produce the part and maintain an effective quality control program. Some PMA-seeking company officials think the process is far too demanding. One complained it took three years of processing after submitting all required data to the FAA.

CHOOSING THE BEST PARTS SOURCE FOR YOU

Despite all the apparent questions and concerns, there are some fairly simple guidelines the aircraft owner can follow. Above all, the single most important step you can take toward assuring optimum benefit for both your checking account and your airplane's condition is to work with an established, reputable repair facility. There is no doubt you can save some money by using PMA parts, but refer to the manufacturer's warranty first. If it says OEM parts only, then perhaps you are better off waiting until the warranty expires before using anything else. Never use unauthorized parts—it voids warranties and when it comes to having maintenance done, shop around for the best price, but be cautious of the lowest bidder.

PART 1
Powerplant

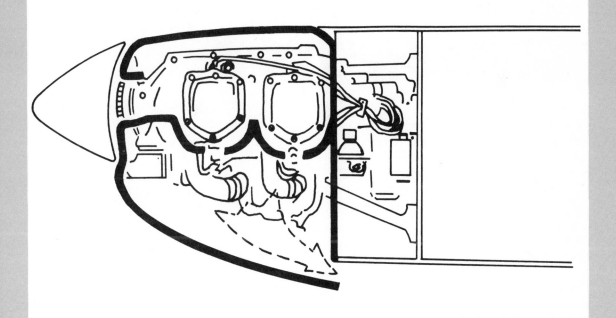

3

Keep on Top of Your Powerplant's Pulse

Your aircraft engine is very similar to the human heart. It thrives on proper diet and exercise, and the way it is treated has a significant effect on its long-term health. Preventive maintenance starts with a proper preflight inspection and encompasses all aspects of engine operation, both on the ground or in the air. Refer to the end of this chapter for a brief listing of terms on aircraft engines.

When conducting a preflight inspection, there are several engine considerations. The cooling air pathway should be checked carefully, including air inlet, air pathway under cowling and the air exit area or, in some aircraft, the cowl flaps.

Oil quantity, of course, is an extremely important item on the preflight check. Pilots should check their aircraft's Pilot's Operating Handbook (POH) to determine recommended oil levels that might differ for cross-country and local flying. Don't forget that oil grade requirements can differ with the season. SAE 50 is commonly used when surface temperatures are expected to be 40 degrees Fahrenheit and above, while SAE 30 (or 10W30) may be necessary when temperatures are colder.

In addition to using the proper oil in cold weather, it is a good idea to pull the prop through by hand at least six revolutions to loosen up the oil. When temperatures drop to less than 20 degrees Fahrenheit and the airplane is not kept in a heated hangar, preheating the engine may be necessary. Normal oil pressure should come up within 60 seconds after start. If pressure does come into the green arc but fluctuates, the oil is still too cold and causing cavitation of the oil pump. Shut down the engine and preheat before another start attempt.

While under the cowling, check general engine security. A visual inspection will turn up most oil and fuel leaks, while a tug at the engine mounts will attest to their integrity. If the airplane is a rental, take a look at the valve covers. The most recent compression check is often written in chalk there and is a good indication of the general health of the engine (more on compression checks later).

Find a good reliable way to determine fuel quantity, such as a fuel dipstick. Fuel gauges, notoriously inaccurate, should not be trusted. Assuring the proper type of fuel also is important; using fuel with a higher octane rating than necessary on a regular basis can cause excessive lead fouling, and too low a rating could result in detonation. The key here is the fuel color and you can check it by draining some fuel into a plastic see-through cup. While checking the color, also look at the quality of the fuel. It should be clean, evenly colored and transparent, without sediment, dirt, or water. Any presence of water is a problem; it is distinctly visible and collects at the bottom of the plastic cup—keep draining and taking samples until there is no longer a trace.

The first rule of good operating practice is always to follow the POH. Whenever faced with a conflict between the handbook and some shortcut or rule of thumb, stick with the POH. However, there are some operational guidelines that generally apply.

ENGINE OPERATIONAL GUIDELINES

Cowl flaps should be open during ground operations, the most difficult cooling situation for the engine. For this reason, run-ups should be done into the wind to maximize cooling airflow through the cowling. Always keep run-ups as brief as possible, using full rich mixture and the appropriate rpm (revolutions per minute). If you have a constant-speed prop, make sure to exercise it as outlined in the POH; this is especially important during the winter.

Don't Baby Your Engine

Follow the takeoff checklist! In general, use full rich mixture unless a high density altitude exists. In that case, lean to recover power lost from an overly rich mixture. Don't baby your engine! Normally aspirated engines have been designed to use full rated power on takeoff. Taking off with less-than-rated power leads to shorter TBOs (time between overhaul). In some aircraft, such as the Cessna 172Q, the recommendation is to maintain full power throughout the climb, while in others, 75 percent power is recommended. In either case, slightly increase climb speed on hot days—obstructions permitting—to get maximum cooling airflow through the cowling.

While the cruise phase of flight presents the lowest apparent engine workload, the pilot must be careful to avoid complacency. Seldom does an engine just fail. Usually, there are indications long before a problem arises. The vigilant pilot will use a regular scan of engine instruments and compare previous readings with current ones. Deviations should be noted and trends watched so intelligent decisions can be made early enough to allow safe, corrective action.

Generally, long descents with the engine at idle should be avoided. The engine cools too rapidly, causing thermal shock and high engine wear. Start descents far enough out to allow for a little power to keep engine instruments in the "green." Make the mixture richer gradually during descent until you are full rich for landing, except in high density-altitude conditions. In this situation, land with the mixture leaned to produce maximum power in the event of a go-around.

Proper Engine Leaning Technique

Whenever engine power is 75 percent or less or during high density-altitude airport operations, you should lean the engine. For taxi, lean at 1000 rpm until rpm peaks; then enrich slightly. During takeoff, use full throttle and lean until you reach maximum rpm with a fixed-pitch prop, then enrich it slightly. If you have a constant-speed prop but a carburetor-equipped engine, lean until the engine runs smooth. Fuel-injected engines should be leaned to the fuel flow setting in the POH.

The application of carburetor heat introduces warmer, less dense air into the engine, which has the effect of enriching the mixture. Using carburetor heat might mandate leaning for maximum performance. Use caution when applying carburetor heat. Some engines require it only when operating in known icing conditions, whereas others call for it whenever power falls outside of the tachometer's green arc. Finally, the pilot always should lean the mixture when operating in excess of 5000 feet (some POHs specify 3000 feet).

Pilots whose aircraft do not have exhaust gas temperature gauges, fuel flow meters, or other more sophisticated instrumentation must lean according to the "rough engine" method. This may be used with any prop as long as the engine has a float-type carburetor. First, set the throttle to the recommended power setting. Slowly lean the mixture until roughness occurs, then enrich it until the engine first runs smoothly. This will give the best economy setting.

ENGINE TROUBLESHOOTING

On a hot day, when you attempt to start a fuel-injected engine that is still warm, it is common for it to come to life only to die swiftly. When the aircraft is parked after flight, residual engine heat "cooks" the fuel lines and metering devices under the cowling. This causes the fuel to expand, which forces it back into the fuel tank, leaving only vapor in the lines. Because a little fuel remains in the lines, the engine roars briefly to life; but the fuel pump is incapable of moving enough vapor to keep it running, so the engine quickly dies.

To avoid this problem, before attempting the start, you should purge the fuel lines by placing the mixture control at cutoff, the throttle at full open (some throttle linkages prevent high pump action with throttle retarded) and turning on the auxiliary fuel pump to high pressure for about 20 seconds. This procedure pumps cool, fresh fuel through the lines, purging the vapor and cooling the system. The fuel return system routes first the vapor, then the fuel back to the tank, leaving the lines filled with fresh fuel. To start the engine, turn off the fuel pump, place the mixture at full rich, open the throttle partially and engage the starter; the engine should start easily.

Rough engine idle could be due to either an excessively lean or rich mixture (that would respond to adjustment) or to a mechanical problem (that should be referred to an airframe and powerplant mechanic). Possible culprits are an induction air leak, improper fuel pressure, bad compression on one or more cylinders, fouled plugs, ignition system problems, or plugged injector nozzle(s).

If the tachometer indicates an excessively low ground-idle rpm, check the carburetor heat or the prop control setting. Other possible problems are a restricted air inlet or governor out of adjustment.

If your engine consumes too much oil, one problem might be the use of the wrong type of oil. Oil type is tough to determine once it is added, so never allow anyone else to add oil to your airplane. Other problems that can cause high oil consumption include worn valve guides and bad or improperly seated rings. The obvious problem in the case of a low oil pressure indication is insufficient oil quantity. More difficult problems might include failure of the pressure-relief valve or a clogged oil pump inlet.

When an engine will not develop rated power, improper use of carburetor heat and mixture top the list of likely causes. Too low a fuel grade runs a close third. If all three check out, there are several other possibilities such as insufficient air induction or fuel flow, low cylinder compression, or incomplete ignition.

Detonation

Engines are most susceptible to detonation at high power settings, particularly if combined with improper leaning. Excessive temperature can cause the fuel/air mixture within the engine cylinders to detonate explosively. This causes a sharp, excessive pressure rise accompanied by a distinct metallic knock (which, unlike automobiles, is seldom audible in an airplane). In addition, there also is a significant temperature increase in the combustion product gases. This temperature rise causes the fuel/air mixture to expand, less fuel burns, and engine power decreases. Another reason for the power loss is the piston's inability to accelerate rapidly enough to convert the unusually high pressure spike into power.

Perhaps the most serious aspect of detonation is its insidiousness. Cracked pistons, burned valves and catastrophic engine failures show up only in the most severe cases. Light to medium detonation might not be noticeable at all in the cockpit, but it can still lead to piston, ring, and cylinder-head damage over time.

Causes of detonation are, typically, too low an octane fuel, excessively hot cylinder head temperatures (CHT), hot fuel/air mixture, excessively lean mixture, and high intake manifold pressure. Provided you have the correct fuel, detonation usually can be stopped by enriching the mixture, making shallower climbs to increase cooling airflow, selecting full-open cowl flaps, and reducing power. If none of these measures solve the problem, you should terminate the flight and seek the help of an A&P.

Preignition

Often confused with detonation, preignition results when the fuel/air mixture is ignited prior to spark plug discharge. The symptoms are similar to, though not as severe as, detonation—engine roughness, backfiring, high CHT, and a loss of power. Preignition can be caused by a hot spot in the cylinder, often a result of a carbon buildup on the cylinder head or spark plugs. It is also possible for a hot spot to develop if valve edges are ground too fine; the thinness causes the valve edge to glow, which in turn, ignites the fuel/air mixture prematurely. As with detonation, significant damage can result such as cracked pistons or valves, so it is important to reduce CHT as quickly as possible. This is done by enriching the mixture, reducing power, and maximizing cooling airflow with higher climb airspeeds and wide open cowl flaps.

PREVENTIVE MAINTENANCE

The best preventive maintenance is to use the engine in normal operations on a regular basis. Gaskets, seals and O-rings need frequent lubrication to stay in condition; long periods of inactivity lead to oil leaks. With time, oil thins and evaporates, allowing the moisture in the air to coat the cylinders and begin the rusting process. Changing the oil as recommended and flying the airplane at least once a week should take care of the worst situations.

If average humidity is below 70 percent, you need only fly the airplane once every two weeks. To be beneficial, it is necessary to actually fly the airplane at cruise for at least 30 minutes each time. Anything less provides insufficient time for the oil-entrained moisture to dry out. Merely doing an occasional run-up does more harm than good. It causes a dramatic temperature change in the engine that when shut down, causes water to condense inside the cylinders, leading to rust and corrosion. Because ground operations in general aren't always very good for an airplane, it shouldn't be difficult to imagine that they are pure poison as an airplane's sole weekly exercise.

While the FAA only requires private aircraft to undergo an annual inspection, next to regular use, the 100-hour inspection is the best preventive maintenance. Engines don't just quit; you'll see symptoms long before anything really nasty happens, and the 100-hour inspection promotes opportunity to look for such telltale signs.

One of the prime indicators of improper operation and engine health are spark plugs. Every 100 hours, the plugs should be removed and checked; normal plugs will have a sort of brownish-gray color. There are three basic problems that this check can turn up: fuel fouling, lead fouling, and oil fouling.

Fuel fouling is indicated by sooty, black deposits, commonly a result of not leaning during high altitude operation (excessively rich mixture). Other conditions that can lead to fuel fouling are too rich an idle mixture, excessive ground operation, frequent power-off descents at full rich, and too low an operating temperature.

Lead fouling, indicated by gray deposits on the plugs, is normal in small amounts. Large buildups mandate frequent plug cleaning and generally are the result of too high an octane fuel. Black, wet deposits, particularly on the bottom plugs, indicate oil fouling. Accompanied by high oil consumption in a high-time engine, this could mean an engine overhaul is due. You will probably subsequently find excessive cylinder-wall wear, worn valve guides, or even a broken piston ring.

The air filter should also be changed at least every 100 hours, more often in high dust or smoky environments. Unlike its automotive counterpart, the aircraft filter doesn't carry dirt well and quickly leads to loss of power, excessively rich mixtures, fouled spark plugs, carbon buildup in cylinders, and even shortened TBOs.

The compression check, a relatively simple test, is a true indication of the engine's ability to produce its rated power. Unlike its automotive counterpart, an aircraft engine compression check compares cylinder pressure against a known pressure, typically 80 psi (pounds per square inch). The FAA requires that the check be conducted by or under the supervision of an A&P for several reasons, not the least of which is the potential danger of being struck by the prop when pressure is introduced.

All cylinders have some air leakage. Therefore you will never get the perfect reading of 80/80. The question is how much leakage and where it occurs. Readings of 75/80 indicate a pretty healthy engine. The FAA says leakage of 25 percent or more of the input pressure means trouble. Because the industry standard input pressure is 80, that means 60/80 and lower is cause for concern. There are three places where the air can leak: past the intake valve, the exhaust valve, and the piston into the crankcase. By listening at the air-induction inlet, exhaust pipe, and crankcase breather cap, the A&P can determine which of the three are the culprit by the sound of rushing air.

In all three cases, however, there still is hope. Bad valves could be the result of a bit of carbon that is preventing complete closure. The mechanic, by giving the rocker valve an educated whack called "staking," hopefully can dislodge the carbon. A subsequent normal reading means you have just saved a bundle of money, but continued low readings indicate a bad valve.

In the case of air blowing by the piston, the problem may be that the engine has cooled too much and the oil has drained away from the cylinder wall, which reduces the airtight integrity. Run up the engine again, and perform another compression check—this might make a difference. If there is no change in the pressure reading, you have a real problem. Readings for all cylinders should be within 5 psi of each other, indicating generally uniform wear. Readings below 60/80 on all cylinders or more than 5 psi difference between one cylinder and the rest require further investigation.

There is one last line of defense before you actually have to remove the cylinders. Your A&P can use an instrument called a *borescope* that permits an inexpensive, visual inspection of cylinder walls, rings, and the top of the piston without disassembly of the cylinder.

The 50-hour inspection—another good investment of time and money—can be performed by the pilot. It calls for a thorough preflight inspection and then a careful security check of all visible systems, such as ignition, fuel/induction, cooling, lubrication, and exhaust. The main thrust is to assure that everything is tight, free of damage, and leak-free, with no excessive wear or indications of heat damage. Finally, the oil should be drained and replaced. Oil loses its effectiveness with time, making oil changes and clean oil filters extremely important to engine health and longevity.

In many cases, the history of an airplane is as important as its symptoms. For instance, if there has been a steadily increasing magneto drop over time, the problem could be old plugs, but a sudden increase could indicate ignition harness or magneto trouble.

Similarly, a steady increase in oil consumption over a long period is a sign of normal wear, leading slowly toward an engine overhaul. If, on the other hand, there is a sudden, dramatic increase, that points to a more pressing problem. Hopefully, a visual inspection will turn up an oil leak that is easily fixed. If not, do a compression check for valve, ring or valve-seat trouble—you might even suspect improper mixing of two different types of oil.

Normal wear over time increases tolerances between moving parts: vibration and noise will eventually appear. Such irritants, if accompanied by a significant increase in oil consumption, can indicate time for overhaul is at hand. Localizing such noises and vibrations can be of great help to the mechanic. In addition, it is helpful to know under what speed, power, and aerodynamic configuration the noise and vibration occurs.

Not only is the health history of the engine important, but so is the operational history. It is of immense value to the A&P to know how the airplane has been operated. Examples of operating conditions that have a significant effect on the life expectancy of the engine are: excessive high power operations, such as a towplane; regular operation from unimproved surfaces, particularly in dusty conditions; and any non-standard operating procedures, such as routine takeoffs with less-than-rated power.

Every airplane owner will eventually be faced with the question of overhaul. As your engine approaches TBO, there are several options, all with very predictable results. It is worth noting that no airplane escapes this ''moment of truth'' and the owner who flies ''cheap'' without an hourly allotment for engine replacement will have a far greater moment of truth than the one who has set money aside for the contingency. In addition to operating the airplane in accordance with the POH, the prudent owner should conduct 50 and 100-hour inspections, that have been proven to extend TBO and cut down costs in the long run.

TBO is a commonly misunderstood term—it differs from engine to engine. For instance, TBO for the Lycoming O-235 (used in Cessna 152s) is rated as 2000 hours. The IO-360 (used in Piper Arrows) is 1800 hours. The Continental IO-520 (used in the

Beechcraft A36 Bonanza and 58 Baron) is rated at 1700 hours. The manufacturer *recommends* the time between overhauls, but that does not mean an overhaul must automatically be done when the engine reaches that "magic" number of hours. The same engine, depending on its maintenance and operational history, may require an engine overhaul significantly earlier than, or even later than, its TBO. However, it has been shown that going well beyond TBO usually produces a more expensive overhaul when the job finally gets done.

Faced with a sickly or lethargic engine, you might have other options besides a total overhaul. If cylinders give a bad compression check, your A&P may advise only a top overhaul. In that procedure, the cylinders are removed and reworked but the crankcase is left untouched. However, a top overhaul never replaces a major overhaul. If the problem were to occur near TBO, it would be foolish to do a top overhaul, only to have to do a major overhaul a short time later. The best course of action probably would be to do the major early and learn from the experience.

Another option is a factory remanufactured engine. In an overhaul, the engine is brought up to manufacturer's specifications for overhauled engines, but the engine's total time continues. For instance, when an engine is overhauled at 1800 hours, you will get it back with 1800 hours in the engine logbook. If you get a remanufactured engine, you will get the airplane back with zero hours, because the engine meets the specifications for a new engine. Only the original manufacturer can remanufacture an engine. For all legal and practical purposes, it is a new engine—in fact, it is unlikely you will get your old engine back. Such a procedure is necessarily more expensive than just an overhaul, but less expensive than installing a brand new engine. For more information on engine overhaul and remanufactured engines, see Chapter 10, Overhauling Your Thoughts On Engine Overhauls.

GLOSSARY

brake horsepower—Horsepower produced by the engine minus losses due to friction, exhaust, and cooling.

detonation—Explosive, near-instantaneous release of heat energy that occurs when a fuel/air mixture reaches its critical temperature and pressure.

direct drive—Propeller bolted to and turned at the same speed as the crankshaft without reduction gearing.

factory remanufactured engine—Engine is completely disassembled and rebuilt by original manufacturer to new tolerances and zero time.

major overhaul—Complete engine disassembly, inspection, and overhaul to manufacturer's specification. Total engine time in the logbook continues.

normally aspirated—Engine rated power lacks ability to maintain sea-level-rated power at altitude.

preignition—Premature ignition of fuel/air mixture in cylinder prior to spark plug discharge.

rated power—Maximum continuous horsepower output when operated at specified rpm and manifold pressure.

time between overhaul (TBO)—Manufacturer-recommended time period to first major overhaul—assumes compliance with recommended operating procedures, maintenance, and inspections.

top overhaul—Entails removal of cylinders, deglazing cylinder walls, installation of new piston rings, and touch-up of valves and valve seats.

4

Engine Instruments Prophesy Performance

The fact that life might have once been simpler does not mean it was necessarily better. Aircraft engine instruments are an excellent example. Since the first tenuous flight, pilots have had some method of checking the well-being of their engines. But the crude instrumentation of by-gone days often left much to be desired. Now, most modern engine instruments are both reliable and useful, provided the pilot understands exactly what is being measured and how.

TEMPERATURE INSTRUMENTS

As temperature increases, liquids and metals expand, though at different rates. By welding together dissimilar metals, coiling them, anchoring one end to an instrument case and attaching the other to an indicator, you have a bimetallic (or solid) thermometer. Light aircraft commonly use this type of outside air temperature gauge; it usually is fitted right through the window or incorporated into the cabin air vent.

Another common non-electrical temperature measuring device uses the vapor method. A gas-filled, sealed bulb and an expandable (Bourdon) tube are connected to an indicator. The bulb is located where the temperature is to be measured; the pressure inside of it varies with the temperature, causing the Bourdon tube to expand and contract, thus moving the indicator.

More sophisticated electrical temperature indicators are of two basic types. *Variable resistance* temperature indicators are based on the principle that a metal's resistance to current flow varies with temperature. When a small, fixed dc voltage is applied to the sensor, some percentage of that voltage, which is determined by the amount of temperature-induced current resistance, passes through the sensor to the indicator. This type of instrument frequently is used to measure outside air temperature, cylinder-head temperature (CHT) and oil temperature. The obvious drawback is that it requires a source of dc voltage.

The *voltage-generation* temperature indicator is based on the principle that certain dissimilar metals that are welded together in a loop produce a low dc voltage proportional to the temperature difference between the two ends of the loop (FIG. 4-1). The *thermocouple* (sensor) is composed of a *measuring junction* (where the loop is joined at the engine) and a *reference*, or *cold junction* (inside the instrument case). A compensating spring automatically adjusts for cabin-temperature variations that might affect the reference end of the loop. Because metals with a very high temperature tolerance may be used, this system becomes ideal for measuring the 1500-degree Celsius exhaust gas temperatures of the reciprocating engine, without requiring an electrical source.

Exhaust Gas Temperature Gauge

The amount of heat produced by the chemical reaction of combustion varies with the fuel/air ratio. If accurately measured, combustion heat is an important diagnostic tool for the pilot. In 1962, Alcor, Inc. introduced and subsequently patented the concept of using exhaust gas temperature (EGT) as an aid to proper mixture control. CHT can be a good combustion problem indicator but lacks the accuracy and directness necessary for precise mixture leaning (FIG. 4-2). In their *Guide To In-Flight Combustion Analysis*, Alcor claims EGT is the most accurate method of determining engine status by measuring combustion heat directly and virtually immediately. In aircraft with fixed-pitch propellers, the rpm or "engine roughness" technique for leaning is used commonly. However, constant-speed propellers prevent detection of rpm variance. There is a more compelling reason for leaning toward the EGT method, even with the fixed-pitch prop—the virtually indistinguishable difference between maximum power and maximum economy settings can be as little as .02 pounds of fuel per pound of air.

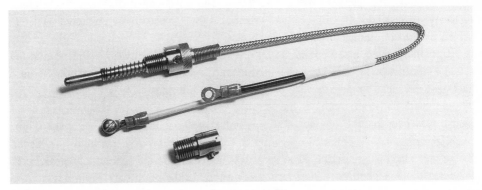

Fig. 4-1. *Cylinder head temp gauge bayonet probe.* Photo courtesy of Alcor, Inc.

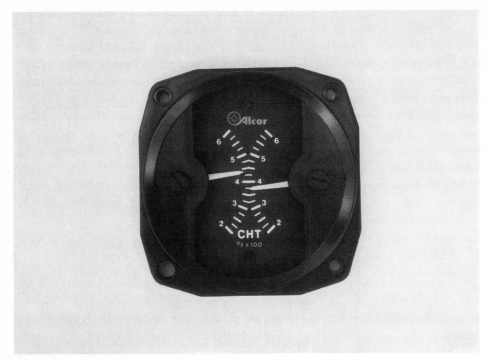

Fig. 4-2. *Twin engine cylinder head temp gauge.* Photo courtesy of Alcor, Inc.

A reciprocating engine can operate across a wide fuel/air mixture range—from .045 pounds of fuel per pound of air at lean misfire to .14 pounds of fuel per pound of air at rich misfire. It is at peak EGT that combustion takes place and the maximum number of oxygen atoms and fuel molecules combine, producing the most efficient cruise condition. Engine operation at mixture settings leaner than the setting that produces peak EGT can lead to cylinder and piston overheating, damage, and finally catastrophic failure. Operating on the rich side of peak is less cause for concern but does produce lead-fouled plugs, costly engine deposits, and increased fuel consumption. The single-probe EGT system, common on many singles and light twins, actually measures the leanest-running cylinder as determined by the engine manufacturer.

In carburetor-equipped engines, fuel distribution differs between cylinders and the actual cylinder experiencing the leanest mixture varies with conditions and altitude. Because an excessively lean mixture is very harmful and the pilot has no way of knowing if the cylinder with the probe really is the leanest, a safety margin must be used to prevent inadvertent over-leaning of one of the cylinders without an EGT probe. Therefore, manufacturers may recommend operating at best (maximum) power which is achieved by enriching the mixture until EGT is about 100 degrees Fahrenheit cooler than peak EGT. While safe, this technique is inefficient and costly. As the pilot leans the mixture from full rich, airspeed increases up to 100 degrees Fahrenheit rich of peak (best power). If leaning is continued to peak EGT, the airspeed begins to decrease slightly, but range and fuel economy increase 15 percent, a significant advantage! Clearly, leaning to peak EGT is desirable but generally not feasible with the single-probe unit.

The proper technique for leaning with a single-probe EGT (FIG. 4-3) is applicable to aircraft equipped with either fixed- or constant-speed props. Beginning at full rich, the pilot should slowly lean the mixture while watching the EGT gauge. As the mixture becomes leaner, EGT increases until the indicator peaks and then reverses. At that point, the pilot should enrich the mixture until it again peaks, stopping 100 degrees Fahrenheit on the rich side.

The multiple-probe system displays an EGT for each cylinder that permits actual determination of the leanest cylinder for existing conditions (FIG. 4-4). The initially higher cost of a multiple-probe system is more than offset by fuel savings and reduced cylinder and piston maintenance. To lean, the pilot first must set cruise power for the appropriate altitude and then rotate the EGT selector knob to see individual cylinder temperatures. After determining which has the highest EGT, the pilot slowly leans to peak EGT on that cylinder. The very act of leaning will change which is the leanest cylinder, so a re-check of all cylinders is mandated. The pilot re-checks each cylinder reading and slightly enriches the mixture while monitoring the EGT gauge. If EGT drops, the cylinder was running on the rich side of peak, which is fine. The mixture should then be returned to its original setting. If the EGT rises, the cylinder is operating on the lean side of peak, and running leaner than the original reference cylinder. The mixture is then reset to peak EGT for

Fig. 4-3. *Single probe exhaust gas temperature gauge.* Photo courtesy of Alcor, Inc.

Fig. 4-4. *Multi-probe combustion analyzer.* Photo courtesy of Alcor, Inc.

this cylinder and the process continued until no other cylinder EGT increases are detected. While tedious at first, practice greatly reduces the time involved, and the benefits are well worth the effort. The process must be repeated with changes in power or altitude.

The EGT gauge is divided into 25-degree Fahrenheit increments, with a major mark every 100 degrees—it measures only relative temperature, not absolute. During preflights, check the thermocouple probes and stainless steel clamps that are approximately 3 inches from the exhaust manifold flange of each cylinder. It also is important to check the integrity of the wire leads.

In addition to proper leaning, the EGT gauge indicates numerous inflight problems. A decrease in both EGT and CHT indicates probable induction system blockage—perhaps icing. A decrease in peak EGT and an increase in CHT indicate detonation. A rapid increase in EGT to off-scale and a rise in CHT indicate preignition. Some systems simultaneously display the EGT of each cylinder, allowing the pilot to keep a constant vigilance.

Graphic Engine Monitor

Insight Instrument Corporation of Buffalo, New York, has gone a step further with its Graphic Engine Monitor which simultaneously displays EGT and CHT for all cylinders. In addition, it automatically finds peak EGT for the leanest cylinder and alerts the pilot to gradual or sudden rise in any one or all cylinder's EGT. It also is an excellent diagnostic tool for problems in ignition, fuel injection, improper fuel use, fuel distribution, and engine failure verification in multi-engine aircraft.

Cylinder Head Temperature Gauge

The CHT gauge (FIG. 4-2), an excellent indicator of how hot the piston, cylinder, and rings are operating, indicates the actual temperature of what the *manufacturer* considers to be the engine's hottest cylinder head. A major concern for any pilot is overheating these parts, which at the least will shorten their lives and will eventually cause catastrophic failure. Operation at high indicated engine temperatures should be avoided, because the hottest cylinder (which varies with conditions and altitude) might not be the cylinder actually displayed. Excessively high CHT can cause detonation, engine damage, and failure. Low CHT coupled with high power can lead to damaged rings and pistons and scuffed cylinder walls.

Causes for excessively high temperature include too lean a mixture, dirty fuel-injector nozzles, an induction system leak, taxiing with cowl flaps closed, extended high power climbs (especially at low speeds), climbs during hot, ambient temperatures, idling the engine with excessively flat prop pitch, and blocked cooling-air pathways or missing/broken cooling baffles.

The two types of CHT probes are *spark plug gasket* and *bayonette*. In the latter, the element is embedded into a special well in the cylinder head and is the more accurate. The CHT gauge is actually a milliammeter with a scale typically calibrated from −50 degrees to +300 degrees Celsius.

Oil Temperature

The oil temperature gauge probe is located where the oil enters the engine. Whether measured electrically or mechanically, oil temperature is displayed on an indicator that is divided into four ranges.

There are two red lines that define the maximum and minimum permissible oil temperature, a green normal operating range, and a yellow cautionary range. The latter indicates a potential overheating hazard, which is an area of concern when using high viscosity oil in low temperature operating conditions.

Before takeoff oil temperature should be in the green. If the temperature never rises into the green range even after a suitable warm-up period, it probably is due to a bad instrument. A takeoff can be accomplished, provided the engine does not hesitate during full throttle application. Any hesitation should dictate an abort and further investigation.

If an excessively high temperature develops during climbout, reduce power and level off to increase airspeed (cooling airflow) to restore the temperature to normal range. A series of such short climbs is known as "step climbing." An excessively cold indication means insufficient lubrication for the engine and could result in dangerous power surges at high power settings.

PRESSURE GAUGES

Two pressure gauges are of primary importance to the pilot. These are the manifold and oil pressure gauges.

Manifold Pressure Gauge

The manifold pressure (m.p.) gauge senses the absolute pressure in the engine intake manifold and displays it on a gauge that is typically calibrated from about 10 to 30 inches of mercury (inches Hg). Twin-engine aircraft typically have only one instrument, but with two superimposed pointers—one for each engine.

In aircraft with fixed-pitch propellers, the tachometer is sufficient to set power, but with a constant-speed prop, the rpm remains constant (within limits) while the throttle controls m.p. Normally, the engine turns the propeller, but if m.p. falls below the green arc, the windmilling *prop* begins to drive the *engine*.

Equally hard on the powerplant is an abrupt throttle control reduction. Therefore, simulated engine failures should be accomplished by cutting the mixture with the throttle open. This allows the manifold pressure to remain high, which assures the combustion chamber stays filled with air—a natural shock absorber for the moving pistons.

Power developed is proportional to the amount of fuel burned, which is based on mass airflow to the cylinders. Airflow is difficult to measure, so *intake absolute pressure* is used as the method of measurement. This is measured at the point just prior to entering the intake valve.

For instance, if you check the m.p. gauge when the engine is shut down, it should read the ambient air pressure (29.92 inches Hg at sea level is standard). With the engine at idle, the m.p. will be very low (for example, 6.15 inches Hg) because the pistons demand more mixture than the carburetor allows, creating a lower-than-atmospheric pressure. At high power, the m.p. will be 26 or 27 inches Hg. Normally aspirated engines never reach atmospheric pressure when they are running, and as altitude increases, the m.p. decreases.

When the pilot selects a higher m.p. with the throttle, the result is an increase in the fuel/air mixture entering the cylinder on each intake stroke. Supercharged engines have ambient air compressed before entering the intake manifold and are capable—especially at sea level—of producing manifold pressure several times greater than ambient. Generally, supercharged engines cannot use full throttle on takeoff or at low altitudes due to the potential for over-pressuring or "overboosting" the cylinders.

According to Avco Lycoming, momentary overboosts are acceptable, but 3 inches Hg for five seconds should not be exceeded. An overboost of 5 inches for 10 seconds indicates the engine should be inspected for possible damage, and overboosts of up to 10 inches anytime mandate a teardown for detailed inspection. Anything more than 10 inches requires an overhaul and crankshaft replacement. A good rule of thumb is to never allow m.p. to be greater than rpm by more than a factor of four—m.p. 24 inches Hg/2000 rpm, 26 inches Hg/2200 rpm, and so on.

Occasionally, manifold pressure gauges behave erratically due to moisture condensation in the gauge line. To solve this problem, most manufacturers put a purge valve between the manifold pressure line and the atmosphere. Pressing the purge button opens the valve, and the higher ambient air pressure enters and forces the water into the cylinder.

Oil Pressure Gauge

Oil pressure, indicated in pounds per square inch (psi), is measured at the outlet of the engine-driven oil pump. Like the oil temperature gauge, the oil pressure gauge employs two red lines that indicate the maximum and minimum permissible pressure, a green arc that shows the normal operating range, and a yellow arc that indicates a cautionary range of potential hazard due to cold-start overpressure and engine-idle low pressure.

Normally, the pressure should be in the green arc within 30 seconds of engine start (or slightly longer if it's very cold outside). During cold start, there might be an indication of excessive oil pressure, but it should decrease when the oil warms up. If not, the engine should be shut down to check the pressure valve for proper setting, and the oil grade for compatibility with ambient temperature. Excessively high pressure leads to oil system failure. Engine power must be minimized until the pressure is correct and stabilized. Prior to takeoff, the pilot always should make one last check of the oil pressure.

In flight, a fluctuating oil pressure indication usually means a malfunctioning gauge. However, it is possible the thermostatic bypass valve is not properly seated. While this is not a significant problem, a cautious eye should be kept on oil pressure and temperature to be safe. Follow up with a checkup by a qualified mechanic as soon as practical. A low oil pressure indication in flight could be the result of several problems, the most common ones being insufficient oil, excessive blow-by past piston rings, and oil leaks. Other causes can be a clogged oil-pressure relief valve or incorrect setting, high oil temperature caused by improper oil grade or quantity, cooling air obstructed to oil cooler, or a dirty oil pickup screen that restricts oil flow to the pump inlet. Because these instruments are basically reliable, if not particularly accurate, any indication of zero oil pressure should be taken seriously and warrants an immediate landing. It might be of some comfort to know that while not recommended for normal operations, most engines can develop oil pressure with as little as two to three quarts of oil. Excessive oil pressure is probably an instrument malfunction, but if it continues to remain high, a landing as soon as practical for a checkup is advised.

TACHOMETER

The tachometer, similar to its automotive counterpart, simply measures engine crankshaft revolutions per minute (rpm). In fixed-pitch propeller aircraft, it is the reference used to set engine power. Aircraft with constant-speed props use the tachometer to set the desired propeller rpm for the condition of flight.

The instrument's range markings are fairly standard among aircraft (with rpm limits varying somewhat), however there are two important variations. In some aircraft, the yellow arc has a time restriction. For example, the Cessna 210N pilot's operating handbook (POH) only permits operations in the yellow arc for 5 minutes at maximum power. The other variation is an occasional, narrow red arc located within the green arc. Operations within the red arc are not permitted except to pass through when increasing or decreasing rpm. The purpose of the red arc is usually to prevent a harmonic or resonant vibration that can lead to structural fatigue and failure. Therefore, due to the uniqueness of markings, tachometers are not interchangeable between different aircraft models.

There are three basic types of tachometer: *magnetic drag*, *remotely driven* and *electronic*. The magnetic drag type is a first cousin to the automotive speedometer. One end of a flexible cable is attached to an engine-driven gear which turns at one-half the engine rpm. The other end rotates a permanent magnet in the instrument case. Around the magnet is a *drag cup* of conductive metal that is "dragged" along as the permanent magnet turns, driving a calibrated pointer. Any indication of inexplicably low power or the inability to synchronize engines in twins should be reason to suspect the tachometer's accuracy if there are no other indications of engine malfunction.

The remotely driven tachometer uses a three-phase, engine-driven ac generator that drives a synchronous motor located inside the instrument. The motor, similar to the drag-cup type, turns a magnet that operates an indicator needle. The primary advantage of this type of system is that there is no mechanical cable. Also, rather than having to rely on voltage to turn the synchronous motor, it is controlled by frequency which is far more stable, making the tachometer accurate through a wide rpm range.

Some aircraft have the more sophisticated electronic tachometer that utilizes a set of special breaker points (they have no ignition function). The tach senses the opening and closing rate of the points and displays it as rpm. Multi-engine aircraft commonly have a synchroscope added to the dual tachometers. This helps trim both engines for identical power settings and aids in synchronizing engines to avoid prop "beat"—a source of both pilot and airframe fatigue.

FUEL INDICATORS

It goes without saying that fuel gauges are an important part of the instrument system.

Fuel Quantity Indicator

The fuel quantity indicator displays fuel remaining for use in computing flying time remaining. For light aircraft, it is very similar to its automotive counterpart—a float-type system. During preflight, it is important to compare the instrument reading visually with the actual tank level. Again, just prior to takeoff, double check fuel quantity and selector position.

The float-type system consists of two devices—a tank (transmitter) unit and an indicator. The tank unit measures the volume of fuel by means of a float that rides on the fuel surface. An arm connects the float to a potentiometer, and as the tank empties, the arm moves across the potentiometer which in turn varies the amount of voltage sent to a remote indicator. The indicator is calibrated to translate different voltages into the appropriate number of gallons remaining.

Pilots have learned to be distrustful of the fuel quantity indicator. For instance, it is sensitive to electrical system voltage fluctuations that can cause erroneous fuel quantity readings. An even bigger problem is due to the many variations in tank design. Fuel tanks that are integral to the wing have a tendency to twist, turn, rise, and fall in an effort to take advantage of any free space where fuel might be stored. This results in float fluctuations. Finally, there is the problem that airplanes just don't sit still; they pitch up, down, yaw, bounce, and roll. The fuel float indicator's efforts are not unlike trying to determine sea level with a cork that is bobbing on the North Atlantic in winter. There are methods of reducing the problem, such as fuel tank baffles, but fluid level is not the best way to determine fuel quantity. Unfortunately, the more accurate methods are very expensive and are left to larger aircraft.

Fuel Flowmeter

As recently as 15 years ago, the fuel flowmeter was restricted primarily to the realm of large aircraft and jets. Now, many general aviation pilots are able to glance at a flowmeter to ascertain the rate of fuel moving from tank to engine (FIG. 4-5). Flowmeters in aircraft

Fig. 4-5. *Tru-flow 1 fuel flow indicator.* Photo courtesy of Alcor, Inc.

with fuel injection actually measure the pressure across a fuel injection nozzle. From one point of view, this approach makes sense because the pressure drop across an orifice is proportional to the fuel flowing through it and the gauge can be calibrated conveniently in gallons per hour. This method has one significant drawback, a plugged nozzle means a fuel flow decrease and a nozzle pressure increase. The gauge interprets this situation as an increase in flow, giving the pilot erroneous information that is opposite to the actual condition!

Aircraft with pressure carburetors use a hinged, spring-loaded plate called a *dynamic hinged transmitter*. The plate partially obstructs the fuel line causing the flow to push against the plate as it passes—the greater the flow, the more the plate is displaced.

The plate pivots on a rotating shaft connected to the transmitter, which in turn, electrically drives the cockpit indicator. The indicator is usually calibrated to show percent of horsepower and fuel flow in gallons per hour. Because there is some variability in the method employed to lean an engine using a fuel flowmeter, it is important to read the aircraft's POH and follow the procedure outlined there.

Engine instruments have one significant similarity to flight instruments—they should be looked at as a whole rather than individually. No one instrument gives a complete picture; a danger indication on one can be evaluated intelligently only after checking the others.

5

How to Troubleshoot Lubricating Systems

Lubricating fluids, namely oil, serve many varied purposes in reciprocating engines. One of the main purposes of oil is to reduce friction—to fill in the microscopic peaks and valleys on the surface of metal. Oil holds metal surfaces apart so the relative movement is actually between two layers of oil. This sliding effect greatly reduces friction and extends the life of the metals.

It is the *viscosity*, or fluid friction factor of the oil that determines how effectively it does its job. Rated in Saybolt Second Universal (SSU), viscosity is determined by calculating the number of seconds it takes 60 cubic centimeters to flow through a calibrated orifice at 210 degrees Fahrenheit. If it takes 65 seconds, it's SSU is Aviation 65. There are also military and SAE rating systems: Aviation 65 is also known as AN 1065 and SAE 30.

The clearance between moving parts determines what viscosity oil should be used. Proper viscosity assures the oil doesn't separate, allowing excessive friction. Other important considerations are, *pour point*, which is the lowest temperature the oil will pour, and *flash point*, the lowest temperature that will cause a momentary flash without sustaining combustion if a small flame is put next to the surface.

Oil also acts as a coolant when it comes in contact with high temperature engine parts near the combustion chambers and areas of friction. Some of the heat is transferred to the oil, which in turn transfers it to the outside air as it travels through the oil cooler.

A third purpose of oil is its ability to cleanse. Oil gathers up particulate matter (such as water, dirt, dust, and flakes of metal and carbon) as it travels through the engine and holds them in suspension. Eventually, the oil passes through the system filter, which traps the contaminants but allows the filtered oil to re-enter the cycle through the engine.

The fourth function of oil is to prevent rust and corrosion. As an engine cools after use, moisture condenses onto the cylinder walls and other engine parts. This moisture and other contaminants lead to internal rust and corrosion. Oil coats these surfaces, preventing moisture and contaminants from actually contacting them.

The last major use of oil is to seal and cushion. Oil helps the piston rings form a seal against the cylinder wall, permitting maximum compression within the cylinder. It also cushions the shock of the moving parts, particularly the valves.

MINERAL OIL

Mineral oil in one form or another has been in use for years; the Wright Brothers used "A-Mobiloil" mineral oil in their early motors. Even though today's mineral oil (that meets military specification MIL-L-6028B) is a well-established, common aircraft engine lubricant, it does have some important drawbacks. When aerated at high temperatures, especially after engine shut-down, oxidation takes place (the formation of carbon deposits). Even at temperatures of 150 degrees and lower, the combination of water vapor, lead compounds, and partially burned fuel tend to "cook" into sludge. This gooey mass clogs filters and can even damage engine bearings.

The short-lived metallic-ash detergent oil was mineral oil with an ash-forming additive (metallic salts of barium and calcium). Initially, it appeared to be the answer to problems associated with mineral oil. It decreased the tendency for oxidation, reduced spark plug fouling to a minimum, lowered the tendency for preignition, and had minimal effect on the combustion process while it facilitated engine cleaning action. The latter was considered especially noteworthy; as the oil travelled through the engine, it removed carbon deposits and sludge. Unfortunately, this type of oil was a disaster disguised as a blessing. The loosened deposits ended up clogging filters and passages and causing general mayhem within the engine. Hence, metallic-ash detergent oil is no longer used in aircraft engines.

On the other hand, ashless-dispersant (AD) oil (meeting MIL-L-22851) has practically taken over the reciprocating engine aircraft market. AD oil has done away with the carbon-forming problems of mineral oil without adding the ash deposit problem of detergent oil—it has a non-metallic polymeric additive. The dispersant additive causes particulate matter to repel each other, preventing sludge. At the same time, the dispersant holds the separated matter in suspension until trapped by the filter. Originally, there was concern that the free-floating particles would act as an abrasive, forming a sort of flowing sandpaper that would wear out parts as it flowed by them. Experience has shown this to be quite the contrary. AD is such a good lubricant that many manufacturers require a new-engine break-in period using mineral oil.

SYNTHETIC OILS

With reciprocating engines operating at higher temperatures than ever before, yet being subjected to varying environments, new types of lubricant have been proposed to satisfy the new needs. Synthetic oil is an attempt to solve the problem of large temperature variations. For instance, a synthetic oil can have the same viscosity at −20 degrees Fahrenheit as a non-synthetic AD oil does at 0 degrees Fahrenheit. Because synthetic oil has a lower internal friction than petroleum-based oils, it has excellent lubricating qualities at very low temperatures. In fact, the observant pilot would notice a 3 to 5 psi lower operating pressure than petroleum oil. Engines using synthetic oil have started without preheating in temperatures as low as −40 degrees Fahrenheit. While there is definitely the potential to eliminate the restriction of preheating, manufacturers still tend to take a wait-and-see attitude by adhering to preheat requirements. Nevertheless, it certainly means greatly reduced preheats, and perhaps the best part is no longer having to drain the oil just because the climate changes: synthetic oil is honestly an all-weather oil.

There are still other advantages to synthetic oil. It produces less oxidation at high temperature and has better wear characteristics than straight mineral oil (about the same as AD), allowing a longer time between oil changes. Probably the best benefit to the occasional pilot/owner is that it adheres to metal better than other types of oil, certainly for weeks and even months! This translates into longer engine life, because it protects the cylinder walls from corrosion and provides instant lubrication on start-up, even for planes that aren't flown regularly. There are disadvantages though. It has a strong tendency to soften rubber and resin products, so you have to be very careful about spillage and leakage. It is also much more expensive than the other types of oil, and while the extended oil-change period tends to compensate for the added expense, a leaky oil system can run into big money.

In research conducted at the prestigious University of Illinois' Institute of Aviation from August of 1979 to January of 1983, multi-viscosity oils were compared to their single-grade counterparts. The multi-viscosity oils were Shell 15W-50 and Gulf 15W-50, both semi-synthetic, ashless dispersant oils, and Phillips 20W-50 ashless dispersant with a straight mineral-base stock.

It was found that the cleanliness and wear of all engines using multi-viscosity oil was the same as those engines using single-grade, ashless dispersants. However, Weldon Garrelts, who headed the research effort, concluded that multi-viscosity oils were superior over single grade in several important ways. He discovered the engines were capable of dependable cold weather starts to 0 degrees Fahrenheit without preheat or noticeable internal engine damage. These oils eliminated the need for seasonal oil grade change. In addition, the average oil consumption was lower for all engines using multi-viscosity oil, and the oil itself performed and retained its properties better. Garrelts did point out that the results do not invalidate the manufacturer's recommendations concerning oil, but he feels engine manufacturers should rethink their standard operating procedures in light of the newer oils.

It is a common misconception that you can't mix different brands of oil. Within the basic categories, all oils are compatible. All ashless dispersants meeting MIL-L-22851 are compatible with each other; they are also compatible with straight mineral oil. If however the engine is high time and has always used straight mineral oil, changing to AD oil might not be as effective as with a lower time engine. If you are planning to switch to a synthetic from either AD or straight mineral oil, then you should drain and flush the system as per manufacturer's recommendation.

AIRCRAFT OIL SYSTEMS

Modern, light aircraft use a wet-sump system. Oil is stored in the sump of the engine and is drawn out through a suction tube by the oil pump. The constant displacement, gear-type pump is the most common in light aircraft (FIG. 5-1). Each time the engine-driven pump rotates, a fixed amount of oil is moved. A pressure relief valve maintains constant system pressure as pump speed varies. The pump has two spur gears meshed together: one is driven by the engine, the other follows. At the inlet side of the pump, the teeth unmesh, causing the cavity volume to increase. This draws oil into the pump, where it fills the spaces between the teeth and is carried around. At the outlet side, the gears mesh, causing cavity volume to decrease which forces the oil out of the pump. Here, in the close

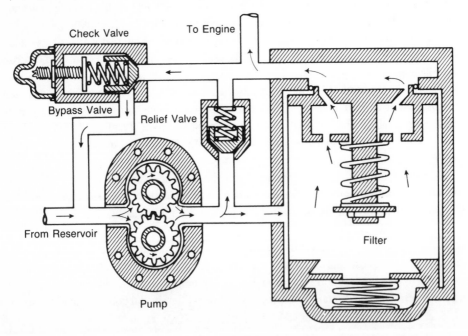

Fig. 5-1. *Constant displacement, engine-driven oil pump.*

quarters of the meshing teeth, is one area where metal chips and other oil contamination can lead to trouble. The source for the oil pressure gauge is tapped off the pump outlet. To prevent gauge fluctuation and minimize oil loss if the line is broken, the hole is very small (approximately $\frac{3}{16}$ths of an inch). The potential to clog such a small hole, or most oil passages, with sludge and other particulate matter is high, so a filtration system is employed.

In addition to clogging oil passages, solid contaminants and sludge can cause significant wear and damage to bearings, rings, cylinder walls, and pump vanes. Typically, a full-flow filter is used, forcing all oil to pass through the filter each time it circulates. The most common filter used in general aviation is the *semi-depth*, which is a long, pleated sheet of resin-impregnated fibers (FIG. 5-2). This sheet is rolled up around a steel core and put inside either a metal spin on container or into a housing that is integral to the engine.

To prevent oil system failure should the filter become clogged, a bypass valve is installed that reroutes oil around the filter. It is worth noting that the pilot will have no indication that the filter is being bypassed; contaminated oil will continue to flow through the engine until the next oil change, or until sufficient damage is done to draw attention to itself. The oil and filter should be changed routinely to minimize this possibility.

A spring-loaded relief valve, downstream from the pump, is used to maintain constant system pressure as the pump speed varies with the engine. If the pump outlet pressure

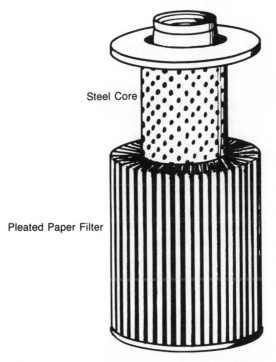

Steel Core

Pleated Paper Filter

Fig. 5-2. *Aircraft oil filter.*

is less than spring pressure, oil continues through the system; if it is greater, the spring is displaced and oil is rerouted back to the inlet side of the pump, causing system pressure to reduce. This process happens so rapidly that fluctuations are not noticeable on the pressure gauge. An adjustable screw varies relief valve spring tension to permit system pressure calibration.

As oil travels through the engine, heat generated by combustion and friction is transferred to it. The oil must then be cooled before returning to the hot sections of the engine. An oil cooler (radiator) is used as a heat exchanger between hot oil and outside ram air. A thermostatic valve routes hot oil through the core of the heat exchanger for cooling or bypasses cold oil around the core. The sensor for the pilot's oil temperature gauge is located where the oil enters the engine, after the cooler. Some gauges are electrically powered while others are mechanical.

Cooled oil is either sprayed or injected onto the crankshaft, camshaft, propshaft bearing, accessory drive bearings, cylinder walls, pistons, and various gears and parts. Both constant-speed propeller governors and turbochargers also use system oil. The oil then drains into the sump and the cycle begins again. To check oil quantity, a dipstick marked in quarts is provided; a filler tube is also provided to add oil. It is important to check both prior to every flight to ensure the caps are secure. Low ambient pressure during flight can cause oil to be siphoned out of the engine if either cap is missing.

PREVENTIVE MAINTENANCE

According to Ben Visser, an engineer for Shell Development Company, maintaining an oil system is really a twofold consideration: selection of the proper lubricant and maintaining oil level. Oil quantity should be maintained at the level recommended in the pilot's operating handbook (POH). Visser stresses that this simple, pilot-performed preventive maintenance measure should be done before every flight. To assure accurate measurement, the aircraft should be in a level attitude. The dipstick should be removed, wiped with a clean rag, pushed all the way back into the dipstick tube, then removed again and read.

Cold weather starting requires special consideration. Always follow the manufacturer's recommendations, but a good rule of thumb is to preheat whenever the temperature falls below freezing. In winter, excessive priming washes oil from the cylinder walls with a twofold penalty: accelerated engine wear and potential fire hazard. Always keep the power setting low until the engine warms up, and avoid any abrupt power change prior to normal operating temperature. Oil, congealed by the cold, will sit in the bottom of the engine until the rising temperature thins it out enough for it to flow. Until then, engine wear is excessive. When warm weather returns, be sure to remove winter oil baffles that reduce oil effectiveness because of unnecessary thinning. It is worth noting that even during summer months, engine oil still needs some warm-up time before flight.

It is important to change oil and filter at or before recommended times. And, the dirtier the atmosphere in which you operate, the more frequent the changes should be. According to Visser, there are actually two times that determine when to change oil: engine time and calendar time. It is common for operators to change engine oil every 25 or 50 hours (depending on manufacturer's recommendations). That's fine for aircraft that are frequently flown, however many aircraft might not fly 50 hours a year. If you are an infrequent flier, you should consider calendar time. Oil should be changed at least three times per year, regardless of engine hours. Oil in low-use aircraft will have a difficult time lubricating the engine because of fuel dilution and condensation. Incidentally, automotive oil is not approved for aircraft use, it contains additives that cause problems in aircraft engines.

Oil leaks shouldn't be taken for granted. Most pilots have come to accept oil stains on the underside of an airplane, but engines aren't supposed to leak oil. If there is a puddle under your airplane or a long wet streak on the fuselage, find out why. If it isn't easily traceable, consult your mechanic. Don't ignore oil leaks, they don't go away by themselves. Knowing your airplane is one of the most important operating tips. Every pilot should be familiar with the oil system's normal operating pressure and temperature. Significant deviations are symptomatic of a problem and should not go unheeded. Similarly, never continue engine operation if either temperature or pressure exceeds red line—extensive engine damage follows closely behind! It is a very rare occasion when an engine fails without warning.

Unused engines have shorter life spans than used ones. That might sound contrary to logic, but it is an established fact. An unused engine should be turned over at least every two weeks either by hand or starter. Don't just start it up, run it for a few minutes and shut it down! When you do this, the engine case heats up, but before the oil really gets hot enough to boil out water and acid, the engine is shut down. As a result, moisture condenses on the inside of the engine which leads to corrosion and rust. If you expect your engine to be inactive for a month or more, the engine should be pickled according to manufacturer's recommendations.

When changing oil, remove the filter and cut it open. Inspect the filter element for any contamination. Metal particles could indicate impending engine failure and should be taken seriously. To prevent inadvertent contamination of spin-on filters that are sealed in a can, there is a special cutter that removes the top without allowing any particles to fall into the filter itself. More than one pilot has been aghast to discover small hunks of metal in the filter, only to find out they were the result of cutting open the filter for inspection.

A good insurance policy is *spectrometric laboratory analysis*. An oil sample from your engine is analyzed in a laboratory to identify various contaminants. Because specific metals are used to manufacture specific engine parts, determination of the origin of the metal contaminant is often possible (TABLE 5-1). Spectro, Inc. of Fort Worth, Texas offers information that outlines a very easy method for obtaining oil samples, either at normal

Table 5-1. Sources of Oil Contamination

silicon—A measure of airborne dust and dirt contamination, it usually indicates improper air cleaner service. Excessive dirt and abrasives accelerate engine wear and can greatly increase operating costs.

iron—Indicates wear originating from any and all steel components, such as cylinder walls, rings, shafts, splines, gears, etc. High iron content can indicate corrosion, if the engine has an inactive history. Often, it will clean out with regular usage if cylinders, cam, and lifters are not pitted.

copper—Indicates wear from bearings originating from any and all steel components, such as cylinder walls, rings, shafts, splines, and gears.

aluminum—Indicates piston metal, piston pin plugs, and can confirm airborne dirt.

chromium—Originates from wear of engine parts that have been chromed, primarily compression rings or cylinder walls.

magnesium—Water reacts with magnesium casings and is carried in the oil. Magnesium also may be an oil additive.

silver—Present in the bearing alloys of a limited number of engines such as the Lycoming supercharged engine, radial-engine master rods, and E series Continentals' front main bearing.

nickel—Can indicate wear from certain types of rings, bearings, and valves or turbo shaft.

tin—Indicator of wear from bearings.

lead—In gasoline engines, the main source of lead is from tetraethyl lead contamination.

oil change or between oil changes, and does not require a mechanic. You simply put the sample in the bottle provided and send it to Spectro, Inc. along with a questionnaire you must answer. Ken Morris of Spectro is quick to point out that the questionnaire is as important as the sample in determining engine condition. He says, " . . . an oil sample without engine and oil data results in about 50% guess as to what the wear numbers mean after processing . . . I could tell many stories on this—wear number looked very good, *but* the engine was using a quart per hour and the owner did not report this."

Morris thinks the system works best when oil samples are taken during oil change and are submitted on a regular basis. Rather than a short term evaluation, this establishes an engine history and wear rate that forecasts long-term trends. When inspecting oil, a visible trace of metal in the filter is not always reason for concern. Some occasional metal flecks are normal; it is excessive flecks that point to trouble. Flecks found in the filter should be sent along with the oil sample for determination of origin. The company analyzes the oil sample and sends you a report that details the amount of each contaminant in the sample, what the likely implications are, and follow-up recommendations if appropriate.

When breaking in either a new or overhauled engine, the oil should be changed frequently. There is a high wear rate during the break-in period, and metal particles can imbed in bearings and severely shorten the engine's life. Typically, oil should be changed at 5 hours or after the first flight. Then again at 10 to 12 hours and again at 25 hours.

It is a good idea to obtain oil samples at each change and have them analyzed. Visser points out the absolute necessity of following the manufacturer or rebuilder's recommendations. While many still adhere to the traditional ''break-in'' with straight mineral oil, some manufacturers and rebuilders have switched to the practice of starting right off with AD.

During routine maintenance, it is a good idea to check spark plugs for oil deposits. Oil isn't supposed to be able to get to that part of the engine, so you know you have a problem if there is more than just a trace. Bad piston rings are a likely culprit; they could be cracked or worn. This might not necessarily show up on a compression check, either. If the compression rings are good, the check will be satisfactory but the other rings might be bad, allowing oil to leak past. The only other route for oil to get to the plugs would be through worn valve guides.

To assure engine longevity, the prudent pilot uses the appropriate oil for the engine and operating conditions; maintains sufficient quantity; changes it in accordance with manufacturer's recommendations; maintains an engine oil analysis program; and is aware of what engine instruments are saying. If pressures and temperatures are running at a significantly different level than normal, there is something wrong.

6

How to Troubleshoot
The Cooling System

There are two basic cooling systems for the aircraft reciprocating engine: liquid and air. Of the two, the most efficient is liquid, but in general terms, the lightest and simplest is air cooling.

Initially, the lightest and simplest air-cooling system that was developed is what is technically termed "velocity cooling." All that means is that the engine and its cylinders are sticking out into the airstream, uncovered where the air rushing past can dissipate the heat from combustion. However, velocity cooling was erratic around the cylinders, especially on the aft side where little airflow reached the fins.

As engine power increased over the years, so did the compression ratios, operating speeds, and therefore operating temperatures. Similarly, there was a demand for a reduction in drag due to cooling to help increase aircraft speed. Initially, enclosing the engine inside a cowling greatly decreased its drag. But, ever-increasing demands led to smaller, tighter engine cowlings and reduced frontal areas (see FIG. 6-1). As engine cowlings became more aerodynamically efficient, they also diminished the volume of cooling airflow. Thus, a need for increased cooling efficiency was created and pressure cooling was the logical solution.

ABOUT LIQUID COOLING

Before we launch into the intricacies of pressure cooling, let us take a brief look at liquid cooling and the reason it isn't used for aircraft. By surrounding the cylinders with liquid, heat is transferred by conduction to the water (or glycol, an ethylene compound

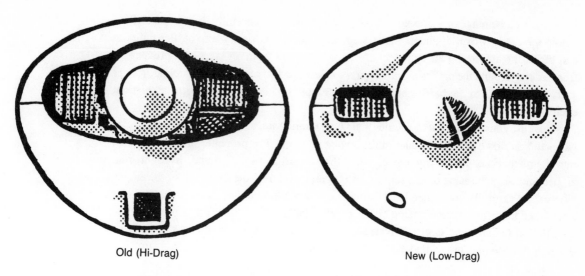

Old (Hi-Drag)   New (Low-Drag)

Fig. 6-1. *Air inlets of older and newer aircraft cowling.*

used as an antifreeze), which is then pumped through a radiator. The radiator's job is to transfer the heat to the air, much in the same way as the cooling fins on air-cooled cylinders. Virtually all automobiles produced (except some Volkswagen models) use this system.

If you want to realize quickly why liquid cooling is not used in aircraft, open up the hood on your car and look underneath. Note the maze of plumbing and the pump and radiator—all points of potential failure in the high-vibration environment of aircraft use. Weight is another factor; liquid-cooled engines are certainly heavier than their air-cooled counterparts. Add to the basic engine the weight of coolant (8+ pounds/gallon), hoses, radiator, and pump.

Liquid-cooled engines have long attracted aircraft designers; such engines can produce nearly double the horsepower from the same cubic-inch displacement! This is possible because the greater heat from higher compression and faster operating speeds can be controlled and dissipated more easily. During World War II, the most powerful and most efficient engines on both sides were all liquid-cooled.

PRESSURE COOLING SYSTEMS

The basic principles of air cooling remain the same in the pressure-cooling system. Cylinders are typically made of chrome-molybdenum steel and the cylinder heads of aluminum alloy. Very thin fins, casts or machined around the outside of both, provide increased cooling surface area for the heat to radiate out into the air. It works the same way a steam-heat radiator works in a house.

The big advantage of a pressure-cooling system is that it carefully directs the airflow within the cowling over cylinders via strategically located baffles. These baffles build up and direct the airflow so all cylinders are cooled equally from all sides. To prevent air from leaking around the baffles and taking a route less conducive to uniform cooling, rubber seals are attached to the baffles and these press against the cowling to form an airtight enclosure.

Imagine yourself to be a molecule of air on a cooling journey around the engine. See FIG. 6-2. You first enter the air inlet of the engine cowling. The propeller pushes you back into the inlet and is your prime mover when the aircraft is on the ground; in flight, ram air accomplishes the same task. From the inlet, you travel over the front, top of the engine where very carefully positioned baffles direct you down around and through the cylinder fins. In addition to assuring engine cooling, these baffles also direct other molecules to cool the oil radiator and engine-driven accessories. Now a hot little molecule, you exit through the opening in the underside of the cowling, rejoining the free airstream.

ENGINE HEAT MANAGEMENT

As an engine's power increases, so does its heat output. Unfortunately, it often is the case that the times of greatest aircraft engine heat output also are the times least conducive to cooling airflow. Situations such as engine run-up, taxi, takeoff, and climb are particular problems for larger engines because the cooling airflow is at its lowest.

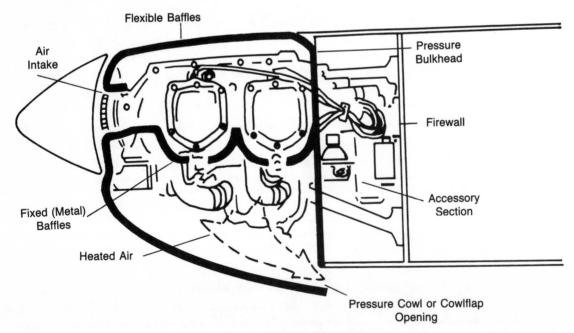

Fig. 6-2. *Engine-cooling airflow.*

To permit precise control of engine temperature, such airplanes have a pilot-controlled door (cowl flap) in the cowling (FIG. 6-3). This door can be adjusted from the cockpit to aerodynamically control air pressure under the cowling. The lower the pressure (cowl flap open), the greater the airflow through the engine compartment, providing maximum cooling for the engine.

As a pilot, your primary engine concern in flight is to make sure it continues to run damage-free. Excess heat is one of your engine's greatest enemies. The most critical heat-related hazard is failure of pistons and/or rings (cylinder heat). Therefore, it stands to reason you should be most concerned about the actual temperature of the cylinder, primarily the cylinder *head*. Unfortunately, most light single-engine aircraft are equipped only with an oil temperature gauge. While it is true that oil contributes significantly to cylinder cooling in addition to lubricating, oil temperature is only an indicator of cylinder temperature. Likewise, the exhaust-gas temperature gauge (EGT) (with sensors installed in the exhaust stacks 4 to 6 inches downstream from the cylinders) senses engine exhaust temperature only. Because it is too coarse to measure cylinder temperature variations, the EGT is primarily a fuel management device.

The primary instrument for engine heating/cooling reference is the cylinder-head temperature (CHT) gauge. The more sophisticated and expensive CHT systems have a temperature probe for each cylinder, but the less expensive single-probe units can still provide the necessary basic information. If your aircraft is equipped with the more accurate

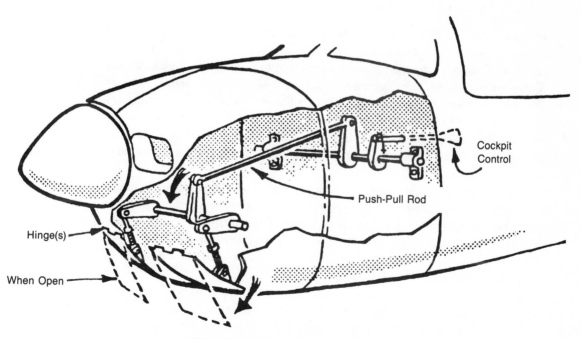

Fig. 6-3. *Pilot-controlled cowl flap.*

bayonet-type multi-probe system, you will see temperature variations between cylinders. Cylinder head temperatures may differ by 100 degrees Fahrenheit in a fuel-injected engine; float-type carburetor engines can vary as much as 150 degrees Fahrenheit between cylinders. This is due primarily to the fact that no cooling system is perfect; there are misaligned baffles and variations in cowling. Also, the position of cylinders relative to location of cowl flaps and placement of engine-driven accessories affects the cooling of individual cylinders. The greater disparity of cylinder temperatures on engines equipped with float carburetors can be traced to the relatively inefficient fuel/air mixture distribution between cylinders.

The first law of thermodynamics states essentially that the rate of heat addition from fuel combustion equals engine power output plus the rate of heat rejection. Because their peak efficiency is only about 30 percent, aircraft engines must dissipate a lot of surplus heat.

Earlier, in tracing the route of a cooling air molecule, it was mentioned that during ground operations, the propeller pushed air back into the cowling, but in flight, the prime mover was ram air pressure. A look at virtually any modern light aircraft shows that the portion of the propeller directly in front of the cooling air inlet has a very low angle of attack and generally is smaller or thicker than the rest of the propeller. Therefore, it is not all that effective at moving air. In addition, the actual air inlet has a fairly small opening. These two conditions significantly decrease the airplane's ability to cool the engine.

OPERATIONAL CONSIDERATIONS

Problems resulting from improper cooling can take two forms: immediate and cumulative. Immediate problems include detonation and preignition. Cumulative problems sneak up on the pilot and can cause permanent, and sometimes catastrophic damage before they are recognized. Here the culprit is not gross overheating of the engine once or twice, but it is rather the cumulative result of continuous use of improper procedures over a long period of time. Each occurrence adds to the problem until there is a major malfunction. The end result is a shorter time between overhauls (TBO) by as much as half the expected life, not to mention scored pistons and cylinders and broken or stuck piston rings.

Taxi and run-ups are critical periods of heat dissipation, and simple preventive measures often are ignored by pilots and mechanics. Always face the airplane into the wind when stationary, especially during engine run-up! A non-moving aircraft has a difficult time getting sufficient airflow through the cowling for cooling, so any help from the wind is beneficial. Avoid unnecessary or long run-ups; they generate excessive heat that is difficult to dissipate; similarly, minimize static high power time.

Never run up an engine under high power with the cowling removed! Frequently practiced by mechanics, this procedure leads to shortened cylinder and piston life, because the cowling is necessary to direct airflow to the lower and back sides of all cylinders and is primarily responsible for getting airflow to the aft cylinders.

Takeoff and climb are also critical periods for engine cooling. An engine should be operated in these phases at full rich mixture, unless the ambient conditions require leaning; excess fuel serves as an additional coolant. It is also imperative that the pilot adhere to the climb speed designated in the pilot's operating handbook (POH). The specified speed is designed to provide the best rate of climb while considering engine cooling. Lower speeds require higher angles of attack which translate into reduced cooling airflow and decreased TBO. A common problem associated with trainer aircraft (or any aircraft used by numerous low experience pilots) is that the aft cylinders tend to burn out early (often significantly below TBO) because of poor climb-out technique. It is very important to get a reliable A&P to check the engine for such premature wear before purchasing a used aircraft.

Generally, cruise and descent are not considered to be cooling problems. However, one thing to watch out for is cooling the engine too rapidly during descent, particularly in larger aircraft. Shifting quickly from cruise to idle power while in a descent can lead to uneven cooling of the engine (thermal shock) and possible failure. The best technique is to plan ahead and begin descending far enough away from your destination to permit a power-on descent.

Engines requiring more precise temperature control are equipped with cowl flaps. The pilot monitors the CHT and opens or closes these flaps as necessary to assure proper engine cooling. Generally, cowl flaps are open during engine start, run-up, taxi, takeoff, and initial climb. They are closed after level-off and remain so during cruise, en route climbs, descents, and landing. If the ambient temperature is particularly hot, the CHT might reflect the need for partial opening of cowl flaps during en route climb or cruise.

Pilots transitioning from less complex aircraft might be dismayed to discover that cowl flaps are yet another thing to remember. When executing a missed approach, cowl flaps need to be opened after initiating the climb because of increased power and decreased cooling airflow. When doing touch-and-go's, the pilot should open cowl flaps before applying takeoff power.

Frequently, you will see pilots—sometimes encouraged by their instructors—keep the cowl flaps open while doing touch-and-go's. The reasoning might go something like this: ''While it may hurt the engine to forget and leave them closed during takeoff, it doesn't hurt to leave them open. And searching for cowl flaps is an unnecessary distraction when you know you are going to be doing a number of touch-and-go's.'' An efficiency expert might tend to agree, but an educational theorist would not. The problem is that this thinking reinforces not operating the cowl flaps at every takeoff, increasing the probability they will be overlooked in emergency operations. Being a pilot means learning to operate the airplane correctly under all situations, and convenient shortcuts can lead to trouble.

Before shutting down the engine after flight, consider that you are about to end what little cooling airflow the engine is getting. Just because you stop it doesn't mean the engine suddenly cools off. The same laws of cooling still apply; the only difference is that there will no longer be any airflow to help dissipate the heat. The air remaining in the cowling will soon be heated by residual engine heat. This might require as much as several hours to cool to ambient temperature during the summer. Meanwhile, all the fuel lines and metering devices forward of the firewall absorb the high temperature, causing fuel expansion. As fuel expands, it is forced back to the tank, leaving vapor in the fuel lines. This fuel vapor in the lines and metering devices causes problematic hot starts. Therefore, on hot days it is a good idea to avoid high power settings after landing and to idle the engine for a few minutes prior to actual shutdown to promote cooling.

COOLING SYSTEM PREFLIGHT

For the most part, preflighting the cooling system is a matter of looking for the obvious. Baffles should be periodically checked for proper alignment; if missing or broken, they should be replaced. Cylinder and cylinder head cooling fins create the majority of engine cooling, so any problems should be brought to the attention of your mechanic.

Airflow blockage is another common problem that should be considered on every preflight. There are three areas of concern: intake, inside of the cowling, and exhaust. Probably the most common blockage is a bird's nest in the air intake or exhaust area. This can be solved by purchasing and using air intake plugs that completely fill up the air intake area, preventing birds or anything else from getting in.

The exhaust area can be conveniently closed up if the airplane has cowl flaps. Otherwise, make sure to carefully check this area, too, because birds are just as liable to build a nest there. Less obvious is a blockage actually within the cowling. My collection of foreign matter discovered in this area includes red, green, blue, and gray shop towels, an instruction manual for an engine compression tester, the beginning of a bird's nest and a torque wrench.

TROUBLESHOOTING

Abnormally high cylinder head temperatures in flight can be the result of a number of problems. First on the list of culprits would be an excessively lean mixture; correct by enriching the mixture. This tends to happen at high power settings during takeoff and initial climb. Similarly, any operations at higher-than-recommended power settings should be avoided. Operations during excessively high ambient temperatures can also lead to unusually high CHT. Some aircraft POHs list a maximum ambient temperature, but typically this is higher than the average pilot will ever encounter. Missing or misaligned engine baffles, air leaking out of the cowling, and airflow blockage are all factors that can cause high operating temperatures. Even a faulty ignition system can show up as a hot reading on the CHT gauge.

The best plan of attack for unusually high temperatures in flight is to enrich the mixture, open cowl flaps, reduce power, and reduce drag as much as possible to keep up the cooling airflow. If none of those have the desired effect and the engine is above the safe operating temperature as outlined in the POH, land the plane as soon as possible. Meanwhile, keep a sharp eye on the oil temperature gauge. If it, too, begins to rise beyond the maximum limit, you have a real problem on your hands and you should find a place to land immediately.

PREVENTIVE MAINTENANCE

For the most part, preventive maintenance of the cooling system couldn't be simpler. Routinely check system integrity with emphasis on keeping the entire airflow path free of obstructions. Also, routinely check the structural integrity of the baffles and fins. Touch them (wiggle them if you can) to make sure that the baffles are not only in good condition but are also properly aligned.

One method of making sure the baffles are doing their job is to spray paint primer on the inside of the cowling over the baffles where the seals touch the cowl. Every 100 hours, check the inside of the cowling for rub marks where you have painted. A lack of marks indicates the seals are not contacting the cowling and air is flowing around the baffle instead of being properly channeled through to cool the cylinders; this translates into shortened TBO. Perhaps the most important thing to remember about this system is that no system is so simple that it cannot cause a problem if neglected.

7

How to Troubleshoot the Ignition System

The aircraft ignition system provides an electric spark to ignite the compressed fuel/air mixture within each engine cylinder. The Federal Aviation Administration requires each certificated airplane with one or more reciprocating engines to have two magnetos per engine, two spark plugs per cylinder and two wiring harnesses connecting the magnetos to the plugs.

Some magneto designs combine two magnetos within one housing. There have been serious questions about whether dual magnetos within one housing really constitute a redundant system. The two magnetos share the same housing, magneto, and common drive shaft but have independent distribution systems, coils, and points. This gives the distinct advantage of lower weight, reduced maintenance, simpler installation, and increased space availability under the cowling.

Optimum-combustion aircraft engines have four strokes per cycle: intake, compression, power and exhaust. The beginning and end of each stroke coincide with the piston positioned either at top dead center (TDC) or bottom dead center (BDC) within the engine cylinder. Theoretically, the spark plug should fire when the piston is at TDC at the end of the compression stroke, where the expanding combustion gases then push the piston down on its power stroke. In actuality, however, the spark plug fires several degrees (of crankshaft rotation) prior to TDC to take advantage of piston momentum. This permits time for normal combustion flame propagation to increase chamber pressure along with the continued compression stroke. The end result maximizes pressure buildup slightly after TDC.

A magneto is functionally an engine-driven ac (alternating current) generator used to create sufficient voltage to jump the spark plug gap to assure proper fuel/air mixture combustion. The magneto requires no external source of electricity. Instead, it creates its own voltage through electromagnetic induction, which simply is a relative movement between a conductor and a magnetic field.

TYPES OF MAGNETO SYSTEMS

There are two types of magneto ignition systems—high tension and low tension. The high tension system, which is most common, actually generates within the magneto the high voltage necessary to fire the spark plugs. This voltage, which is transmitted through a distributor and wiring harness to the appropriate spark plug, has a tendency to jump from the harness and/or distributor and follow the path of least resistance. Known as "flashover," this is especially a problem in the low pressure and cold atmosphere of high altitudes.

Another problem with the high-tension system results from some leads being longer than others because the plugs are farther away from the magneto. These long leads tend to store energy, releasing it after the normal timed ignition. This second spark is called "capacitance after-firing." High in heat energy, it attacks spark plug electrode materials and causes approximately double the electrode erosion rate. A high-tension system is relatively lower in cost, less complicated to install than its low-tension counterpart and lighter in weight, making it ideal for most general aviation light aircraft.

The low tension system overcomes flashover by transmitting low voltage from the magneto through the harness, which then is stepped up through a transformer near the spark plug. In addition to performing better at high altitudes, the magneto also has greater resistance to inflight moisture problems, operates more efficiently, and the shorter high-tension leads reduce potential voltage loss as a result of the lead's low capacitance.

Magneto timing is critical to engine health and performance; an improperly timed spark can cause significant problems. Timing that is too advanced can lead to power loss, overheating, detonation, and preignition. If timing is too retarded, a significant power loss and increased fuel consumption would be expected.

MAGNETO SYSTEM PREFLIGHT

Checking the condition of magnetos can be accomplished in two ways: grounding check and differential check. Magnetos are connected by a primary wire (P-lead) through the magneto switch to ground. Its purpose is to provide an internal path of least resistance, rendering the magneto inoperative when the switch is turned off. If the P-lead fails, the magneto stays "hot."

The significant drawback of this system is on the ground, where a "hot" mag patiently waits for someone to come along and even slightly turn the propeller, causing the impulse coupling to produce a spark that could make the engine roar to life.

To check for proper grounding of the magneto, set the throttle at idle and rapidly move the mag switch from BOTH to OFF, then back to BOTH. You should hear a momentary engine failure. If there isn't one, then one or both mags are hot and you need to find a mechanic.

Incidentally, P-leads go to specific mags and are not interchangeable. During start-up, there is insufficient piston momentum to overcome normal "advanced" sparking prior to TDC, so the starting spark is "retarded" at or near TDC on the compression stroke. An impulse coupler is used to snap the magneto through its firing position rapidly so it will produce a high energy spark even though the engine is being cranked slowly, and simultaneously it retards the spark. In such a system, it typically is the left magneto that has the impulse coupler, while the right mag is grounded automatically during the start procedure to prevent its *advanced* spark from prematurely igniting the fuel/air mixture and causing the piston to kick back.

The entire system (impulse coupler, retarded left spark, grounded right magneto, and starter motor) is controlled by the starter position on the ignition switch. Switching magneto P-leads will deactivate the impulse coupler and cause the wrong mag to fire in advance of TDC. The probable result will be an engine that is trying to run backwards, with potentially significant damage. When hand propping, never put the switch to BOTH for the same reason. Always set it to the magneto that has the impulse coupler. Even the more expensive systems that use a starter-vibrator instead of an impulse coupler operate functionally the same and require proper mag selection when hand propping.

The differential magneto check, familiar to all pilots, compares one mag to the other. Using the manufacturer's recommended power setting, ground one magneto and note rpm difference from BOTH. Then ground the other, note its difference from BOTH, and compare the difference from the previous reading. The rpm drop is the result of incomplete combustion. Using only one set of spark plugs is not as efficient as using both sets, so there is a loss of engine power and rpm. The manufacturer publishes a maximum-allowable rpm drop from BOTH and a maximum allowable rpm difference between magnetos. If the rpm drops in excess of these figures, ground the aircraft.

Several other possible conditions might be noted during the differential mag check. For example:

- Immediate, excessive rpm drop probably indicates failure of the secondary mag coil, which prevents a sufficient current flow to get the spark to jump across the spark-plug gap.

- Slow, excessive rpm drop indicates a problem with magneto timing. If the engine slowly dies, it again is probably because of a failure of the secondary coil, which prevents sufficient high voltage at the lower rpm used for the check.

- No observable drop in rpm indicates probable P-lead failure.

- An engine failure during the mag check can be traced to probable failure of either the primary coil or the breaker points. *Caution:* Do not switch back to BOTH—

unburned fuel/air mixture can build up in the cylinders and exhaust. Switching to both will ignite the mixture and possibly cause major damage to the cylinder and exhaust. Throttle to idle, then switch ignition to BOTH to prevent damage.

☞ A condenser failure typically shows up as high rpm missing and is not noticeable during low rpm testing. Points will prove to be pitted and burned upon inspection.

It is worth noting that operating the engine on one mag for as little as 30 seconds can lead to fouling of the inoperative plugs.

Moisture, which finds its way into a magneto simply by condensing when the temperature changes, can leave corrosive acids behind. These acids act as carbon trails begging electricity to flash over. It is a good idea to reduce the problem as much as possible by checking for cracks, loose cover plates, or other preventable methods of moisture entry.

Preventive maintenance should include a timing check at 100 hours or annual inspection, whichever comes first. Routine maintenance typically is at 500 hours. At that time, the mechanic checks, among other things, contact point assembly burning and wear, distributor gear carbon-brush wear, and cracking, chipping, and integrity of the impulse coupling shell and hub. A magneto should always be overhauled (or replaced) at engine overhaul or any other time conditions warrant it.

THE SPARK PLUG

The spark plug transfers high voltage current into a cylinder by causing it to jump through the fuel/air mixture between the spark plug's center and ground electrodes, thereby igniting the mixture. If that sounds simple, then consider the conditions: voltages in excess of 18,000 V; wildly fluctuating gas temperatures as high as 3000 degrees Fahrenheit; and greatly varying pressure from partial vacuum up through 2000 psi! The latter is particularly significant, because as compression pressure increases, the magneto voltage required to spark the gap also increases. As if these conditions were not bad enough, the plugs must not allow naturally emitted ignition radiation to interfere with sensitive navigation and communication equipment, and they must do all this reliably for long periods of time. In 1966, R.L. Anderson of the Champion Spark Plug Company calculated that during a 100-hour operating period, any given spark plug is required to ignite approximately 7,000,000 combustion charges!

Several variables are associated with choosing a spark plug. Particularly confusing is a plug's heat rating; *hot* and *cold* are the options (FIG. 7-1). The rating reflects the plug's ability to transfer combustion-chamber heat via the insulator core nose to the cylinder and engine cooling system. Hot plugs transfer heat relatively slowly; therefore the insulator core tends to stay hotter. Cold plugs transfer heat quickly and tend to stay cooler. The correct heat rating assures the plug will operate cool enough to prevent preignition, where the mixture is prematurely ignited because of some "hot spot" within the cylinder, such as a glowing hot portion of the spark plug. The correct heat rating also will assure that

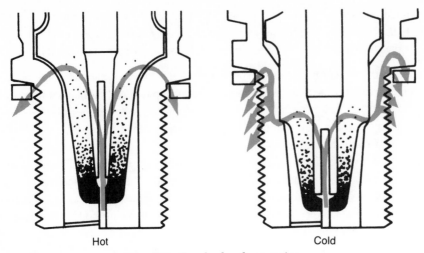

Fig. 7-1. *Spark plug heat rating.*

the plug operates warm enough to resist plug fouling by burning off unwanted contaminants.

Aircraft spark plugs use either *fine-wire* or *massive-core* electrodes. Fine-wire electrode spark plugs tend to be self-cleaning, which greatly reduces the chance for misfiring. The electrodes are actually made of precious metals, typically platinum or iridium, because these metals virtually prevent lead deposits from adhering. The plugs should be cleaned and gapped every 100 hours; if properly maintained, they can last 1700 to 1800 hours! Other benefits include easier starting, greatly reduced incidence of plug icing, and no high altitude flashover. Massive-core plugs should be cleaned and gapped twice as often as fine-wire electrode plugs (every 50 hours), and even then, the average life is in the range of 350 to 400 hours. It doesn't sound like much of a deal until you consider that for the typical light-duty recreational aircraft, these dependable plugs could easily last for several years and can be purchased for half the price of fine-wire plugs.

Resistor plugs reduce the heat energy of capacitance after-firing, which in turn reduces the severity of electrode erosion. Because the problem is associated primarily with high tension magneto systems, resistor plugs will be of little use with a low tension system. Properly used, they cut erosion in half.

The reach of a spark plug is measured as the distance between the shell gasket seat to the end of the shell thread (FIG. 7-2). Reach is determined by the cylinder-head design. The proper reach places the electrode in the best position to ignite the mixture. If the reach is too long, exposed threads easily can become ''hot spots'' that can cause preignition. When replacing plugs, always use a new gasket, because reach includes the thickness of a new gasket under proper torque. However, note that on plugs where thermocouples are used, no gasket is added.

Routine spark plug maintenance should be conducted every 100 hours unless otherwise stated. The plugs should be removed and checked for proper gap, fouling deposits, and

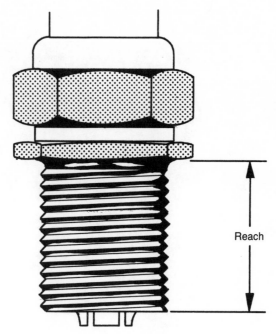

Fig. 7-2. *Spark plug reach.*

general condition. Cracked porcelain usually indicates preignition. Stan Fletcher, Champion Spark Plug Company's supervisor of aviation marketing services, says he considers spark plug electrodes to be ''. . . one of the major indicators of engine health.'' A good mechanic can read an electrode like an open book. Typically a brownish gray color, the electrodes readily identify problems associated with ignition, fuel mixture, and piston-ring wear.

As the electrode gap widens due to erosion from normal service, there is a loss of approximately half the voltage margin between the potential available from the magneto and the voltage required by the spark plug to fire. At some point, the magneto will be unable to supply sufficient voltage to jump the ever-increasing gap. When that happens, the engine begins to misfire. Gap erosion is a normal consequence of operation, but excessive erosion can be an indication of improper fuel metering timing, magneto timing, or plug heat range. Other causes include capacitance after-firing and constant magneto polarity, which occur in engines with an even number of cylinders. In that situation, any given plug always fires with either positive or negative polarity. Positive polarity plugs result in excessive ground electrode wear, while negative polarity results in excessive center electrode wear. To equalize the wear of both polarity and capacitance after-firing, swap plugs after removing and cleaning them so that top and bottom plugs change places and long and short lead plugs change places (FIG. 7-3).

Fouling deposits are the biggest problem in spark plug operation. Carbon fouling, which appears as a dry, dull, black color, is the result of an excessively rich mixture,

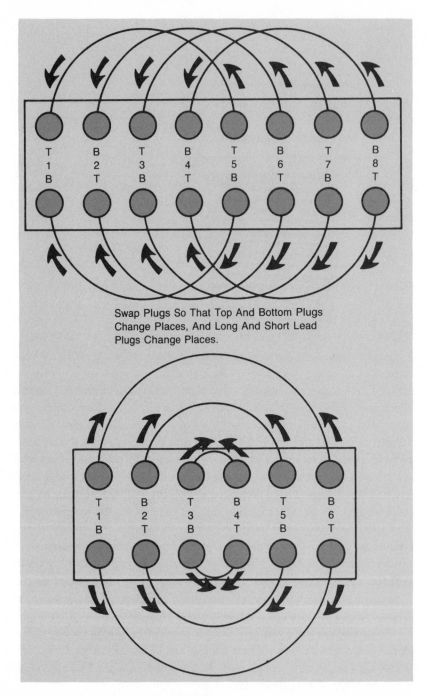

Swap Plugs So That Top And Bottom Plugs
Change Places, And Long And Short Lead
Plugs Change Places.

Fig. 7-3. *Spark plug position changes.*

especially at idle. Spending too much time at idle and on the ground causes significant carbon fouling of the plugs. Another possible cause is a plug with a heat range that is too cold to burn off the combustion deposits. As the carbon builds, risk of misfiring during full power application increases. To reduce the potential for carbon deposits, operate the engine at the highest possible power setting and leanest carburetor setting conducive with safe ground handling conditions. Periodically apply a momentary higher power to cleanse the plugs. Similarly, a faulty carburetor can lead to carbon fouling.

Oil fouling shows up as wet, black, carbon deposits. Mild deposits are common on lower plugs, especially if the aircraft is seldom used, because the oil tends to drain past the rings. New engines also experience oil fouling, indicating the rings haven't properly seated yet; the problem should correct itself in a short time. If oil fouling appears on the upper plugs, there are several possible causes, none of which are very good. They include: worn or broken rings, damaged piston, worn valve guide, sticking valve guide, and faulty ignition supply. It is especially worth noting that oil is electrically conducive, and fouling can lead to misfiring under any power conditions.

Evenly dispersed, fluffy-looking deposits that range from tan to dark brown in color are indicative of lead fouling. It is always present in some limited amount, but excessive lead fouling is typically caused by one of the following: high lead content in the fuel, poor fuel vaporization (can be caused by carburetor air being too cold), operating the engine too cold, or the plug's heat range is too cold. If left unchecked, lead will eventually fill up the end of the plug and cause the plug to operate colder. This leads to misfire at high power settings, carbon collection, and then additional misfire trouble from the carbon!

Pilots seldom are aware of the delicate relationship between dust and spark plug health. Lead oxide, one of the least objectionable spark plug deposits because of its high melting point, considers silica (found in dust) a delicacy. It digests it and forms lead silicate with as little as 3 percent to 5 percent silica ingestion through the air induction system. Lead silicate has a melting point several hundred degrees below that of the original lead oxide. It causes a free-flowing spark plug contaminant, leading to misfiring at *normal* temperatures! The process takes from 20 to 50 hours of engine operation to develop. Therefore, it is critical to frequently clean air filters when operating in dusty environments.

During winter months, if cold air with a relatively high moisture content enters a warm cylinder during shutdown, moisture condenses on the electrodes and freezes. The ice can form a conductive bridge to the electrode and prevent an engine start the next time. The only options are to remove and thaw the spark plugs or apply external preheat to the engine.

Removal and installation of plugs is not as simple as it sounds. Aircraft spark plugs are amazingly delicate, considering the conditions they are required to operate in. Proper tools and techniques are essential. Never install a spark plug that has been dropped; always throw it away immediately. It is virtually a ''given'' that such a plug has been damaged, most probably in the porcelain, but the damage might not be visible.

Installation absolutely requires a torque wrench. Over-torquing will not improve the connection and can damage both the plug and the bushing. A loose plug will not transfer heat properly and can cause preignition when it overheats. Cleaning plugs requires a special tool, so you will probably have to have them cleaned by a mechanic.

When removing the plugs, put them in a tray or other device to make sure you know which plug came from the top or bottom position of which cylinder. Trays designed for this purpose are available, but an egg carton is just as effective if you label the indentations. The threads of the spark plug also should be cleaned before replacing it. Dirty threads cause poor contact between plug, gasket, and engine seat which reduces heat transfer, causes excessive plug temperature and increases the chance of seizing the plug. Many mechanics apply an anti-seize compound to the threads to make removal easier next time. However, there are some cautions. Never use a graphite-based compound, because it acts as an electrical conductor. Instead, use either an authorized anti-seize compound, anti-rust compound, or plain engine oil. When applying, start at least two threads away from the electrodes—otherwise it might run off and short the plug.

Conductive wiring carries electrical current between the magneto and the individual spark plugs and is protected by an insulated sheath known as *shielding*. Designed to protect the wiring from heat, atmospheric conditions, and vibration, this wiring harness also must suppress electrical interference with radio communication and navigation equipment. It is a good idea to periodically check the harness for deterioration from heat, age, and general cleanliness of terminal contact springs and moisture seals (to prevent flashover). If you are doing your own plug cleaning and gapping, don't forget to clean up the ceramic terminal connector sleeves of the wiring harness with acetone, alcohol, or naptha.

The external inspection should include, if possible, a visual inspection of the magneto hold-down nuts. They should be tight and safety-wired. If they are loose, there is a good possibility that the magneto timing will be off. When handling the propeller, always assume the mag is "hot" and never turn the prop unless absolutely necessary, and then only as if it were going to fire to life! Once the engine is started, do a thorough cockpit magneto check, including the ground check. Many pilots prefer doing the ground check as the last item before engine shutdown. The plugs and harness, where visible, should be checked for general security. In addition, the harness also should be inspected for signs of aging, cracking, and chafing against the cowling. When possible, check connections for tightness.

If there was ever an aircraft system that deserved to be given top preflight priority, the ignition system is it. All the fancy equipment, retractable gear, and avionics in the world won't be very helpful if the engine stops running in flight.

8

How to Troubleshoot the Aircraft Fuel System

The purpose of a light aircraft fuel system is to store fuel safely and deliver the correct amount of it at a uniform flow to the carburetor or other fuel control unit. There are two types of systems: *gravity feed* and *pressure*.

Gravity-feed systems rely on the force of gravity to deliver fuel from the tank to the carburetor, which limits them to high-wing aircraft (FIG. 8-1). The Cessna 152 uses a gravity-feed system because it is very simple, relatively inexpensive, and does not require a fuel pump. This type of system is limited to smaller, single-engine aircraft.

The most significant drawback to the gravity-feed system is its tendency to develop vapor lock, a condition where fuel changes from a liquid to a vapor. When the fuel lines are filled with vapor, they are unable to supply sufficient fuel to maintain engine combustion, resulting in a fuel-starved engine. Vapor lock is caused by excessive fuel temperature, high altitude operations, or a combination of the two. Fuel vaporization is most likely the result of shutting down an engine on a hot summer day; when fuel lines under the cowling are exposed to extremely hot temperatures with no cooling airflow. High altitude operation, where there is lowered atmospheric pressure, can also induce vaporization at a lower temperature. The solution to chronic problems of vaporization is to provide positive fuel pressure with a fuel pump.

In the pressure system, the engine-driven pump draws fuel from tanks located anywhere in the airplane and discharges it under positive pressure to the carburetor. This permits greater flexibility in tank utilization and minimizes vapor lock potential. A second, or auxiliary, fuel pump is used for priming, engine start, and as a backup in case of engine-driven pump failure.

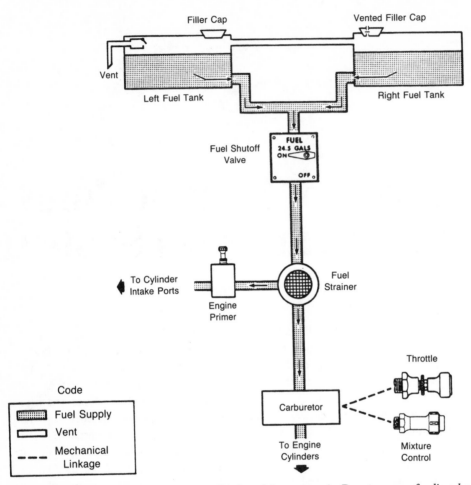

Fig. 8-1. *Cessna 152 fuel system (standard and long range). Due to cross-feeding between fuel tanks, tanks should be topped off after each refueling to ensure maximum capacity.*

As aircraft size increases, so does the fuel system complexity. More sophisticated than the Cessna 152, the Beech M35 Bonanza is designed for longer trips and instrument flying. Its pressure-feed fuel system (FIG. 8-2) is more complicated than the C152's gravity-feed system, plus it has auxiliary fuel tanks.

When a pilot transitions from a single-engine aircraft to a twin, he or she becomes aware of how much more complex the fuel system is. Such aircraft are designed for long range, extensive instrument flight and emergency single-engine operation. The Beech B55

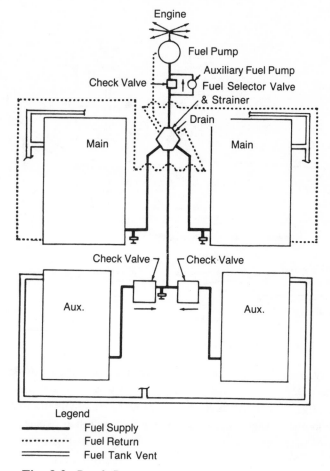

Fig. 8-2. *Beech Bonanza K35 and M35 fuel system schematic.*

Baron, for instance, has a more complex fuel tank and fuel/engine selector arrangement than virtually any single-engine airplane built (FIG. 8-3). The drawback of increased fuel system complexity is the potential for increased fuel mismanagement. In some aircraft it is possible to suffer unstable fuel balance conditions. Each aircraft has its own peculiarities, and good operating practice dictates a thorough understanding of system procedures.

FUEL TANKS

Fuel tanks come in all shapes, sizes, and locations. Integral fuel tanks are permanently built into each wing, while tip tanks are attached to the wing tip and often contain auxiliary fuel. Fuselage tanks can be located almost anywhere in the body of the airplane.

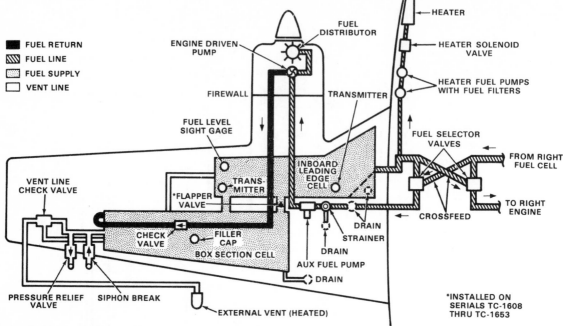

Fig. 8-3. *Beech Baron B55 fuel system schematic.*

Temporary, portable fuel tanks can be located in the cockpit to provide supplementary fuel for extended flights, such as ferrying aircraft over water, but they present considerable fire and noxious gas hazard. All tanks must be protected from vibration—a principle cause of deterioration and leakage—and must be able to handle fuel expansion as a result of heat. Another consideration is surging, a fuel interruption caused by the fuel flowing toward the wing tip during some turning maneuvers. The problem can be reduced by placing baffles inside the fuel tank; this slows lateral fuel movement, providing normal flow during the steepest climb, best angle of glide and all other normally anticipated flight conditions. Tank design also must include a method for the pilot to check for fuel quantity and quality during the preflight.

Most fuel tanks are made of either pre-shaped, riveted aluminum alloy or synthetic rubber. Sealants are used to ward off fuel leaks around the seams of aluminum tanks, but sealants tend to deteriorate and leaking tanks often result. To remedy this problem, the tank can be welded, however this is a costly process that results in a heavy, expensive tank that's susceptible to vibration fatigue. The apparent answer is fuel-resistant, synthetic rubber bladders that are lighter than metal and very flexible. They are fitted easily into available space, relatively easy to replace, and are much less susceptible to vibration fatigue.

Some types of rubber bladders become dry and brittle with age and begin to leak. However, filling the tank after each flight usually prevents this problem. Tanks made of Goodyear BTC-39 synthetic rubber (used in many Cessna, Beech, Rockwell International, and Piper aircraft during the '60s and '70s) developed a different problem—softness. An airworthiness directive (AD) required them to be inspected annually for deterioration; use of BTC-39 synthetic was finally discontinued by Goodyear.

Another problem that affects all rubber fuel bladders is wrinkles. When a synthetic rubber fuel bladder is installed, it is very difficult to get out all of the wrinkles. On the ground, this acts as a water trap, however inflight motion dislodges the water with a high potential for engine failure. This has become such a concern in the big Cessna singles (180, 182, 206, and 210), the company has issued an owner-advisory bulletin stating, among other things, that owners should "gently move and lower the tail to the ground" during preflight, which will hopefully cause the water to dislodge and show up in the preflight fuel sample.

Several experiments, however, have shown that such a procedure can have little effect in removing entrapped water from these Cessna models. Consequently, AD 84-10-01 was issued last year, requiring, among other measures, the installation of additional quick drains in these aircraft and an extensive check of the bladders for wrinkles and their effect on trapping water. If, after compliance with the AD, the bladder still traps more than 3 ounces of water, then an elaborate preflight inspection procedure is required, plus additional annual inspection considerations. It is simpler, and probably safer in the long run, to replace the fuel bladder with a new one.

Other parts of the fuel tank include vents and overflow drains. As the fuel level of a tank decreases, the fuel vent allows air to fill the space. It is important to preflight the vent; if blocked, fuel starvation will stop the engine and possibly collapse the tank. Overflow drains act as safety valves for fuel when it expands as a result of heat. Nevertheless, the prudent pilot should allow for fuel expansion.

Fuel lines, normally made of annealed aluminum alloy or copper, must be of sufficient diameter to allow *double* the required flow rate at takeoff power. They must be protected from excessive vibration, and flexible hosing is required where the lines are actually connected to the engine or airframe. Hoses should be checked for chafing during preflight. While manufacturers try to route fuel lines away from exhaust manifolds and other hot areas, it is not always possible. Asbestos tape serves as heat protection and should also be checked on preflight.

WATER CONTAMINATION

Water can enter a fuel tank three ways. The first way is by the formation of condensation on the inside of partially filled tanks when outside air temperature drops. Topping off tanks after each flight eliminates this possibility. The second way is for water to leak past the fuel filler cap, a problem that exists especially with recessed filler caps that form a cup which holds rain. To minimize this problem, inspect filler caps, seals, and ports every preflight. Also check for fuel stains trailing behind the filler cap; in flight, reduced pressure over the wing causes fuel to stream out of a leaking filler neck. The third way for water to enter a fuel tank is for it to be pumped in with the fuel. Fixed-base operators' fuel trucks and storage tanks suffer from the same problems as airplane fuel tanks, but the trucks have filtering systems built in. Reputable FBOs properly maintain this equipment and train personnel to use it correctly.

A strainer is located between the fuel source and carburetor to trap impurities and water. Quick drains are usually located at the lowest point in each fuel tank—and for the entire fuel system—where water and sediment tend to collect. During preflight, the pilot should check a fuel sample for purity and correct color; any mixing of fuels will cause the color to become clear. Water is heavier than avgas and sinks to the bottom of the cup to form a distinct "bubble," while sediment appears as floating specks. Always use a clear container for collecting the fuel sample so you can inspect it carefully.

FUEL PUMP AND DELIVERY SYSTEM

Pressure systems have two types of fuel pumps: engine-driven and auxiliary. The engine-driven pump runs off the accessory panel at the rear of the engine. Typically an eccentric sliding vane type, it provides a positive displacement with a large volume output of fuel to the carburetor. As with most fuel pumps, the rotor is lubricated by the fuel itself, virtually eliminating the need for any kind of preventive maintenance. The auxiliary fuel pump is required as a backup for the engine-driven pump. Originally, these pumps were hand operated (known as *wobble pumps*), but today they are electric. In addition to their role as backup pumps, they also are used during engine start to build up fuel pressure, during takeoff as a safety margin, at altitudes above 10,000 feet to reduce potential for vapor lock, and during emergency operations.

In most light aircraft, a single pump is sufficient for normal operations, but in larger ones, electric boost pumps often are required. The purpose of the boost pump is to supply fuel under pressure to the engine-driven pump, thus preventing vapor lock. Failure of a boost pump usually means operational limitations imposed on the aircraft, including altitude restrictions and even a drastically reduced useful life of the engine-driven pump.

The purpose of the fuel selector valve is to allow fuel manipulation by the pilot, including tank to engine, tank to tank, and in multiengine aircraft, cross-feeding one engine from the opposite fuel tank. In the event of engine fire, emergency procedures call for the fuel to the engine to be shut off, a procedure that can alone put out the fire. As the pilot's options increase, so do the hazards, therefore several safe operating practices should be adopted. First, never change the fuel selector position just prior to takeoff. Why switch to the unknown? Second, never operate by "feel" alone; always check visually, because some fuel selectors require you to go through the OFF setting when switching from one tank to another. Third, always test all tanks while you still have options; don't deplete one without knowing if the other will work properly. Finally, flight instructors should be aware that when they switch off the fuel selector to simulate an emergency, they sometimes create more than one! A safer procedure is to pull the mixture control.

Both manual and electric primers supply a small amount of fuel to the cylinders for engine starting. The manual type is a single-action piston pump. The pilot pulls out the primer, creating a partial pressure that causes fuel flow into the primer cylinder. When pushed in, the fuel is forced through a primer distributor to either the individual cylinders or the intake manifold, depending on the system. The main advantages of a manual primer are that it is inexpensive and simple. A significant disadvantage is the fire hazard that results from fuel routed through the cabin to the primer. During preflight, take a very close look at the primer itself behind the instrument panel to make sure there is no fuel leak. Under the cowling, look for and check the condition of fine tubing that runs into the cylinders or intake manifold.

The electric primer is a solenoid plunger; it acts like a door, permitting fuel to flow from a source pressurized by the auxiliary fuel pump to the primer distribution system. Because it is controlled by a remote switch in the cabin, you would preflight this system the same way you would a manual engine primer, except you don't have to worry about fuel being in the cabin.

The fuel pressure gauge takes its measurement at the carburetor inlet. There are two types: *bourdon tube* and *autosyn transmitter*. The bourdon tube is a small, coiled tube with a movable, sealed end. A fuel pressure increase causes the tube to unwind, and the end mechanically moves an indicator needle in the gauge. It has the advantage of being simple, inexpensive, and totally self-powered, but like the manual primer, it is a fire hazard because fuel actually is brought into the cabin to operate it.

The autosyn transmitter measures fuel pressure the same as the bourdon tube, but the pressure-sensing device is mounted on the engine side of the firewall with a transmitter armature electrically sending the appropriate indication to the cabin instrument. While it does require electrical power to operate, its significant advantage is that no fuel ever enters the cabin.

REFUELING CONSIDERATIONS

As an aircraft moves through the air, the resulting friction causes it to become negatively charged, much the same as rubbing your shoe on a carpet. Cold, dry weather makes it worse. The Earth, which tends to be positively charged, would cancel the airplane charge when it lands, but tires make very poor conductors of electricity. Along comes the flight line attendant driving the fuel truck. Unless the airplane is grounded to the fuel truck and the Earth, a spark will likely jump from the tank to the hose nozzle, causing an explosion. While most pilots agree that proper grounding procedures should be followed when returning from a flight, many believe these procedures are unnecessary when refueling an aircraft that has not been flying. The very motion of the fuel flowing through the nozzle creates static electricity and can cause a spark between the aircraft and the refueler. Therefore, proper grounding procedures always must be adhered to when refueling.

Another concern during refueling is contamination. It previously was mentioned that buying from a high volume, reliable dealer will greatly reduce this type of problem. That is not always possible, but there is a device that can help. A portable fuel filter made by Pilot's Pal of Anchorage, Alaska temporarily attaches to virtually any kind of fuel nozzle, including a 5-gallon can. Only 3 by 19 inches, it easily stores under a seat; with its cap in place there is no gas odor in the cabin. The unit has a clear portion that permits visual inspection of fuel color and a male ¾-inch garden hose fitting on one end to allow extension hose lengths if necessary. The disposable filter not only removes both water and solids with a flow rate of 10 gpm, but when saturated, it automatically stops the flow, assuring uncontaminated fuel. Pilots who operate in wilderness areas or fly to airports with questionable fuel supplies can hardly go wrong.

Assure your aircraft is fueled properly by always watching the fueling procedure and actually checking the fuel yourself—the fuel truck could have been misfueled! To help prevent misfueling, have all refueling ports distinctly labeled with the fuel type and install the new misfueling-preventive hardware available for most aircraft. This two-part system of preventing incorrect fueling requires the installation of restrictors that decrease the size of the aircraft fuel tank opening. All new aircraft will come factory-equipped with the new fueling ports. The other part of the system, an oversized jet fuel hose spout, is being installed on refuelers by FBOs around the country. The avgas tank retrofit costs about $35 to $40 per tank, it can be done by the pilot, and is available from many FBOs.

TROUBLESHOOTING

If the engine simply refuses to start, the most obvious problem is no fuel—either the tanks are empty or the fuel valve is off. Other, more complicated problems requiring a mechanic's expertise could include inoperative primer, a plugged or ruptured fuel line, improperly set or inoperative carburetor mixture control, and any one of several carburetion problems.

Perhaps even more frustrating is an engine that roars to life only to die suddenly. If the airplane has been flown recently and it is a warm day, one of the most common causes is vapor lock. The pilot should follow the hot-start procedure outlined in the pilot's operating handbook to purge the fuel lines of vapor. Other possible culprits include a clogged fuel vent line (which may or may not be obvious upon inspection), an inoperative engine-driven fuel pump, (which must be checked by a mechanic), a filled fuel strainer (easily cleaned by the pilot), or water in the fuel system (pilot can usually drain).

If the engine runs, but excessive black smoke is coming from the exhaust, you need a mechanic. Probable causes are: the idle or cruise (depending on where it happens) engine mixture setting is too rich; a continuous primer leak into the intake manifold; or any one of several carburetor problems.

9

How Turbochargers Boost Performance

Most pilots know that a normally aspirated engine loses power as the aircraft climbs because air density (the number of oxygen molecules per cubic foot) decreases as altitude increases. Similarly, high ambient air temperatures cause an engine to produce less power, again because of decreased air density. (Warming the air causes the oxygen molecules to spread apart, lowering the density.) So, if engine power is directly proportional to the mass of air and fuel burned in its combustion chambers, it should be a simple matter of increasing the flow of fuel/air to regain lost power.

In a way, it really is that simple. A larger fuel pump will increase the flow of fuel and an air pump (compressor) can pack more molecules into the cylinders. The complexity comes in constructing, attaching, and driving the air pump.

Mechanically driven compressors, called *superchargers* for their act of increasing the density within a specific volume, have been with us nearly as long as the airplane itself. At one time, virtually every transport or military aircraft was equipped with the geared supercharger. Most big radial engines have them, though they require a lot of power to drive. You still can find examples of mechanically driven compressors on diesel trucks and drag-racing cars. Virtually everyone else now uses exhaust-driven turbo-superchargers. Turbo-supercharging, now generally referred to as "turbocharging" by "turbochargers," has also been around a long time. The first experiments with turbochargers on aircraft took place in the early 1920's. Some of the more advanced World War II military airplanes had them: notably the P-47 Thunderbolt, P-38 Lightning fighters, and B-17 bombers . . . all of which could top 30,000 feet! Their advantages are many: lightweight, compact, simple to install, and they're a "free" source of power.

DESIGNED TO MAINTAIN SEA LEVEL POWER

In general aircraft use, the turbocharger's role is not to increase engine power at sea level; rather, it is designed to maintain sea-level power as the aircraft climbs. To do so, the turbocharger must maximize engine volumetric efficiency. *Volumetric efficiency* is the ratio of total cylinder volume to actual displacement. For instance, if the intake stroke of a piston completely filled the cylinder with fuel/air mixture at ambient pressure, it would be 100 percent efficient.

Normally aspirated engines are those that draw in the fuel/air mixture by the suction created on the downstroke of the piston. During the downstroke in a four-stroke cycle, the pressure in the cylinder becomes less than that of the ambient air, so the fuel/air mixture is "sucked" into the cylinder through an intake valve. Such engines are never 100 percent efficient for a number of reasons: at high rpm, the entire operation happens so quickly that the mixture simply doesn't have enough time to fill the cylinder before the intake valve slams shut. Under the best of conditions, the airflow is slowed by obstacles, such as bends in the manifold. Even if 100 percent efficiency were possible, there still is the problem of decreasing air density which accounts for as much as 50 percent loss of power at 12,000 feet.

The solution to some of those problems is the turbocharger. Instead of mechanical linkage connecting the unit to the engine, a free-turning, vaned wheel is placed directly in the engine exhaust stream. This *turbine wheel*, driven by engine exhaust, is connected to the impeller portion of a centrifugal compressor assembly (FIG. 9-1). The impeller directs ambient air to the spinning compressor axis. The air is spun at a very high velocity, forcing it outward and causing increased pressure. This high pressure air then is ducted to the intake manifold of the engine. See FIG. 9-2.

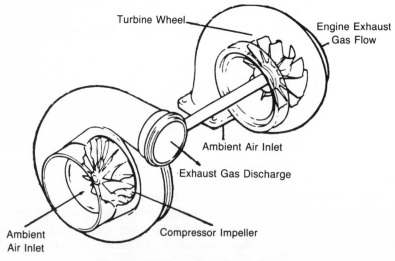

Fig. 9-1. *Basic turbocharger diagram.*

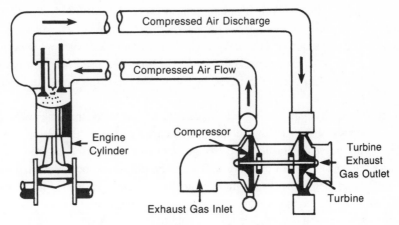

Fig. 9-2. *Turbocharger flow diagram.*

Using engine exhaust to power a turbocharger makes it seem like you're getting something for nothing, but there are disadvantages. One of the major problems is the adverse operating conditions of the turbocharger itself. The turbine and compressor routinely rotate at speeds in excess of 100,000 rpm. The turbine, which drives the entire unit, is subjected to extremely high engine exhaust temperatures.

PROBLEMS OF HIGH ALTITUDE OPERATIONS

There also are special circumstances to consider when operating at the higher altitudes that become available through the use of a turbocharger. Fuel vaporization is one problem. Engine-driven pumps pull fuel to the intake manifold, which at high altitude, invites vapor lock. So, the aircraft must be equipped with tank-mounted boost pumps to feed fuel to the engine-driven pump under positive pressure. Boost-pump failure can cause cavitation, and eventually failure, of the engine-driven pump, as well as vapor lock and engine fuel starvation.

Another problem associated with high altitude operation is the electrical conductivity of the rarefied atmosphere. Magnetos and wiring harnesses require special care and protection to prevent electrical problems that cause engine roughness and possible failure.

MISCONCEPTIONS REGARDING TURBOCHARGERS

A number of misconceptions surround the use of turbochargers. It is commonly believed that turbochargers increase fuel efficiency. They don't, but they do allow you to take advantage of higher true airspeeds and more favorable winds at higher altitudes.

Some operators think the increased fuel/air mixture causes greater stress on the engine. It would seem logical, but actually the opposite is true; a turbocharged engine has less operating stress. Remember, a normally aspirated engine has four piston strokes: intake, compression, power, and exhaust. Compression, power, and exhaust strokes occur in a positive-pressure condition, causing fundamentally the same type of pressure on the piston, rings, and connecting rod. The intake stroke, however, creates suction, which causes a significant change of force on the piston. This force change is transferred to the crankshaft via the piston connecting rods. The faster the engine operates, the worse the effects are of the pressure change.

The turbocharged engine has all positive pressure strokes; when the intake valve opens, air from the turbocharger is pumped into the cylinder as fast as the piston can move downward. This prevents drastic force changes and is easier on the engine.

Despite the consistency of pressure within the cylinders, turbocharged engines have shorter TBOs than their normally aspirated counterparts. One factor used in determining TBO is how hard the engine is worked during an average hour.

With a turbocharged engine, the pilot is able to operate at rated takeoff power for a significantly longer time. Whereas the normally aspirated engine begins to lose power immediately after takeoff, the turbocharged engine routinely operates at a higher percentage of its rated power during a greater part of its lifetime. That means the engine must dissipate more heat over a longer period of time. That greater amount of heat results from higher power settings, in addition to the heat generated by the turbocharger itself. Remember, when air is compressed, it increases in temperature. That hot, compressed air is mixed with fuel and shoved into the engine, and that heat reduces engine life.

CONTROLLING A TURBOCHARGER

From an operational standpoint, a turbocharger needs one additional piece of equipment. If the exhaust gases spin the turbine, which in turn spins the compressor, the system needs a way of controlling the free-wheeling turbine and the resultant air pressure. This is accomplished with a *wastegate*, a damper-like device that controls the amount of exhaust that hits the turbine rotor.

A full-open wastegate directs exhaust straight through the exhaust pipe and overboard, so it doesn't hit the turbine rotor at all and the compressor produces no pressurized airflow. As the wastegate is moved toward the closed position, more and more exhaust is channeled to the rotor, causing it to spin faster and faster.

From a practical standpoint, the wastegate is like a variable slough that controls the amount of water that goes to a waterwheel. The slough can divert water away from the wheel, channel a little to it or direct a great amount of it to the wheel.

Early turbocharger systems required the pilot to control the wastegate directly. It would be open for takeoff, and then as the engine began to lose power in the climb, the pilot would close the wastegate gradually to maintain power. Occasionally, pilots would attempt to take off with the wastegate closed. The resultant ''overboost'' can provide spectacular takeoff and climb performance, but like trying to stuff 10 pounds of potatoes into a 5-pound bag, the resultant excessive manifold pressure would eventually destroy the engine. Similarly, some pilots would forget to re-open the wastegate as the aircraft descended and an overboost would occur. Early turbocharger systems did not compensate for changes in airspeed, pressure, or temperature, resulting in disconcerting fluctuations in manifold pressure (m.p.), called "bootstrapping." As a result, three basic wastegate control systems evolved: fixed, throttle-controlled, and automatic.

The fixed, ground-adjustable wastegate remains in the same position throughout all engine operations. The exhaust flow is split, some of it dumping overboard and some going through the turbine. A very simple system, the pilot controls m.p. at all times with the throttle. The advantages of such a simple system are that it requires minimal maintenance and is relatively low in cost, however, there are some major drawbacks. The most obvious one is the significant loss of potential power in the exhaust dumped overboard. By its very nature, this system has a low critical altitude, though it is better than having no turbocharger at all. Perhaps the most important drawback of the fixed wastegate is that the throttle is the only m.p. control. Therefore, the compressor might produce more pressure than is necessary, thereby subjecting the engine to more heat than is good for it. Finally, the fixed wastegate is susceptible to the nagging problem of bootstrapping.

The next step up is a turbocharger system that has a mechanical linkage connecting the throttle to the wastegate. At low power settings, the throttle opens normally, not affecting the full-open position of the wastegate. Once the throttle reaches a full-open detent, the mechanical linkage then allows further throttle increases in conjunction with automatic closure of the wastegate. This provides some control over the turbocharger, but bootstrapping can still be a problem.

The two types of automatic wastegate controllers are the *pressure-reference* system and the *density-reference* system. The former maintains a selected manifold pressure as set by the throttle. Engine oil pressure, directed by the controller, moves the wastegate as necessary to maintain the appropriate m.p. The obvious advantages of the pressure-reference system are that the m.p. does not need to be adjusted manually during climb, and the maximum m.p. is limited so the pilot does not have to worry about overboosting.

The density-reference system, used on some Lycoming engines, goes one step further. At full throttle, the wastegate is controlled by compressor discharge air. The pilot selects the desired m.p. and the *density* (or *slope*) *controller* tries to hold that density of air, regardless of changes in airspeed, ambient pressure, or temperature. This system can actually increase the m.p. by several inches during a climb, which has startled more than one unsuspecting pilot. Many pilots react to this situation by backing off on the throttle in the mistaken belief that the controller has failed and is putting out too much pressure. At less than full throttle, the turbocharger reverts to being a pressure-reference system and things work pretty much as the pilot expects. The obvious advantage to such a system, when used correctly, is that it is automatically controlled and it compensates for variation in density and airspeed. It almost totally eliminates bootstrapping.

THE PROBLEM WITH HEAT

For all the positive aspects of the turbocharger, it still has the nagging problem of hot, compressed air being pumped into the intake manifold. Not only does heat destroy engines, it also robs them of power. A given turbocharged engine might produce the same m.p. and rpm at 13,000 feet as it does at sea level, but that doesn't mean it's producing the same horsepower. The hotter the air, the less dense it is, and while the turbocharger packs molecules of air together to overcome the decreased density of air that comes with altitude, it is not unusual for 15 to 20 percent of the engine's horsepower to be lost to heat, which means the turbocharger must work that much harder to make up the difference. This in turn creates more heat, causing a vicious circle.

To reduce the extent of the heat problem, some manufacturers put *intercoolers* between the compressor discharge and the intake manifold. The intercooler cools the hot, compressed air before it goes to the engine. Not only does that increase the critical altitude, it also reduces the potential for detonation as a result of overheating. The disadvantages are: increased weight and drag (the intercooler uses ram air), additional system complexity, and added expense.

With a turbocharger system, the manifold pressure gauge is controlled by the throttle, and the tachometer is controlled by the propeller lever. The exhaust-gas temperature (EGT) gauge displays mixture control and the cylinder head temperature (CHT) gauge is controlled by cowl flaps and airspeed. From an operational standpoint, the key to system longevity is heat control.

During takeoff and climb, EGT is controlled by throttle setting. If the EGT gets too high, power must be reduced. During cruise, EGT is a direct result of mixture manipulation; however, rpm also has an effect on EGT. If engine rpm is so fast that the exhaust valve opens before combustion is complete, EGT will be hotter because combustion will take place in the exhaust manifold.

Many pilots think takeoff and climb are the most critical periods of operation for an engine; actually, cruise can be far more critical. It is true that during takeoff, the engine is under full power, but it is also using a rich mixture with the excess fuel acting as an additional coolant. In cruise, despite the lower power setting, a significantly leaner mixture is used, which creates greater potential for overheating. Remember, at cruise, mixture is the most important EGT control.

One of the most overlooked instruments is the cylinder head temperature gauge (CHT). During takeoff and climb, cowl flaps should be open and the proper airspeed should be flown. If the temperature becomes excessively hot with full-open cowl flaps, the pilot must increase airspeed with a shallower climb, so the engine gets a greater flow of cooling ram air. Even cruise might require partially open cowl flaps, especially when operating at economy power settings. Remember, the turbocharger is stuffing hot air into the engine.

Another confusing aspect of this situation is that cylinder cooling is reduced at high altitude. That might not seem logical because it is so cold at higher altitudes, but less dense

76

air has very poor cooling properties. If the pilot cannot keep CHT within an acceptable range at high altitude, he or she may need to request a lower altitude to take advantage of greater air density.

Descent requires consideration, too. Turbocharged engines operate hotter at higher altitudes. This provides a tremendous potential for thermal shock during a descent. Except in an emergency, the pilot should never pull the power back to idle at high altitudes. Plan on making power descents, use flaps and gear if necessary to increase vertical speed, but maintain power. A good rule of thumb is to use 5-inch reductions in m.p., with a couple of minutes between reductions. This allows time for the engine to adjust to the decreasing temperature.

When preflighting the engine, carefully examine the turbocharger and exhaust pipes for cracks. The engine and turbocharger should be checked for loose fittings, oil leaks, cracks, cuts, or holes. Surrounding areas should be inspected for paint blisters or corrosion, indicating excessive heat. Other potential problem areas include clogged air cleaners and clogged engine crankcase breathers. If a preheat is necessary, take sufficient time to ensure the entire engine is heated. The temperature probes might be heated sufficiently to give an erroneous indication in the cockpit while oil in the engine sump and outlying accessories is still congealed.

TROUBLESHOOTING

The single most important rule of thumb in troubleshooting a turbocharger problem is to check the engine first. Far too many units are repaired or replaced only to have the problem reappear immediately. Most turbocharger problems are caused by one of the following: lack of lubrication, foreign object damage, or contamination of lubrication. The most common problem is lack of lubrication, which usually shows up first as bearing failure, wheel rub, seal damage, or shaft breakage. Foreign objects can damage either the turbine blades or the compressor, but in either case, a wheel imbalance at 100,000 rpm can be devastating. Contamination of lubricant causes scored shaft journals and bearings, blocks oil holes, plugs seals, and eventually leads to heavy oil leakage.

There is little troubleshooting that can take place in flight. The system either works or it doesn't. If there is a sudden loss of m.p., the turbocharger is the likely culprit, and you will have just reverted to a normally aspirated engine. That alone is not a significant problem, unless you are over high mountains. The most important question is why did the turbocharger fail? Watch for fire! Carefully monitor pressures and temperatures and watch for a loss of engine oil. Any sign of either fire or loss of oil merits an immediate engine shutdown. One other preventive measure you can take is to listen to the system. Become familiar with the sound of your turbocharger; it has a high whistle or whine. Should that become abnormally loud or shrill, there probably is insufficient bearing clearance, and the unit needs servicing to head off a failure. Use TABLE 9-1 as a guide.

According to Richard Popplewell of Aircraft Accessories of Oklahoma, Inc., the best

Table 9-1. Turbocharger Troubleshooting

Engine lacks power	Black exhaust smoke	Excessive engine oil consumption	Blue exhaust smoke	Turbocharger noisy	Cyclic sound from turbocharger	Oil leak from compressor seal	Oil leak from turbine seal	Cause	Remedy
•	•	•	•			•		Clogged air filter element.	Replace element according to engine manufacturers recommendations.
	•	•	•	•	•	•		Obstructed air intake duct to turbo compressor.	Remove obstruction or replace damaged parts as required.
•	•		•					Obstructed air outlet duct from compressor to intake manifold.	Remove obstruction or replace damaged parts as required.
•	•		•					Obstructed intake manifold.	Refer to engine manufacturers manual and remove obstruction.
			•					Air leak in duct from air cleaner to compressor.	Correct leak by replacing seals or tightening fasteners as required.
•	•	•	•	•				Air leak in duct from compressor to intake manifold.	Correct leak by replacing seals or tightening fasteners as required.
•	•	•	•	•				Air leak at intake manifold to engine joint.	Refer to engine manufacturers manual and replace gaskets or tighten fasteners as required.
•	•	•	•			•		Obstruction in exhaust manifold.	Refer to engine manufacturers manual and remove obstruction.
•	•					•		Obstruction in muffler or exhaust stack.	Remove obstruction or replace faulty components as required.
•	•		•			•		Gas leak in exhaust manifold to engine joint.	Refer to engine manufacturers manual and replace gaskets or tighten fasteners as required.
•	•		•			•		Gas leak in turbine inlet to exhaust manifold joint.	Replace gasket or tighten fasteners as required.
			•					Gas leak in ducting after the turbine outlet.	Refer to engine manufacturers manual and repair leak.
		•	•			•	•	Obstructed turbocharger oil drain line.	Remove obstruction or replace line as required.
		•	•			•	•	Obstructed engine crankcase vent.	Refer to engine manufacturers manual, clear obstruction.
		•	•			•	•	Turbocharger center housing sludged or coked.	Change engine oil and filter, overhaul or replace turbo as required.
•	•							Fuel injection pump or fuel injectors incorrectly adjusted.	Refer to engine manufacturers manual—replace or adjust faulty component(s) as required.
•	•							Engine camshaft timing incorrect.	Refer to engine manufacturers manual and replace worn parts.
•	•	•	•			•	•	Worn engine piston rings or liners (blowby).	Refer to engine manufacturers manual and repair engine as required.
•	•	•	•			•	•	Internal engine problem (valves, pistons).	Refer to engine manufacturers manual and repair engine as required.
•	•	•	•	•	•	•	•	Dirt caked on compressor wheel and/or diffuser vanes.	Clean using a non-caustic cleaner and soft brush. Find and correct source of unfiltered air and change engine oil and oil filter.
•	•	•	•	•		•	•	Damaged turbocharger.	Analyze failed turbocharger, find and correct cause of failure, overhaul or replace turbocharger as required.

preventive maintenance is to catch problems before they become major. "At 100,000 rpm, little problems become big problems fast." Oil leaks, unusual vibrations and sounds should be followed up carefully. Nothing is as effective as good operating procedure. When shutting down the engine, the oil pressure drops to zero, and so does the lubricating capability of the turbocharger. The typical approach and landing are at low m.p., allowing plenty of time for the unit to spool down. But if high m.p. is maintained on the approach or is used during taxi, the turbine will still be spinning after engine shutdown, with no lubrication for it. The prudent pilot should set the engine at idle for several minutes prior to shutdown to assure adequate lubrication and spool-down time.

It is also important to change oil at least at the frequency prescribed in the pilot's operating handbook (POH). The bearings are highly susceptible to dirty or improper oil. Turbochargers should be overhauled at the same time the engine is, at the recommended time between overhaul (TBO). In the long run, this reduces unscheduled downtime and gives the owner a lower operating cost.

Spinning turbocharger parts are machined to within one one-millionth (1/1,000,000) of an inch. It is a good idea to conduct any required maintenance or inspections on time. Assure there is no buildup of exhaust carbon or warpage in the wastegate. Pilots who operate in dusty or smoky air should be especially careful to change the oil frequently and following prescribed inspections. If it sounds as if turbochargers are problematic, then you should realize that at least one manufacturer, Garret AiResearch, has no scheduled periodic maintenance requirement for its turbochargers. If you follow normal engine maintenance and keep the oil clean, there should be no problem at all!

Most modern turbocharged engines have overboost protection. Don't rely on it! Always allow a minimum of 30 seconds warm-up before running up an engine (longer when it's cold); this prevents oil lag. On takeoff, advance the throttle gently to about 25 inches while holding the brakes. Let the turbocharger come up to speed, then release the brakes and gently advance the throttle to the m.p. indicated in the POH for takeoff.

10

Overhauling Your Thoughts on Engine Overhauls

Looming on the horizon of every aircraft in service is engine TBO (time between overhaul). When your mechanic tells you it's time, there are three possible maintenance solutions: replace your engine with a new one, replace it with a rebuilt engine, or overhaul it. Replacement with a new engine can be extremely expensive and is typically not an economically viable solution. To make the best decision, it is necessary to understand the terms related to engine overhauling and rebuilding.

ENGINE OVERHAUL

A very commonly misunderstood engine-related term is TBO. Most pilots expect to get the manufacturer's recommended TBO out of their engine. They don't understand that the emphasis is on *recommended*. Handling and care determine the actual time between overhauls. The pilot who practices proper engine care will almost certainly be rewarded with an actual time between overhaul in excess of the manufacturer's recommended. And yes, you can exceed TBO as long as the engine continues to operate properly! On the other hand, the negligent pilot is likely to run out the engine long before reaching TBO.

According to FAR 43.2(a), an overhauled engine is described as follows.

> (1) "Using methods, techniques, and practices acceptable to the Administrator, it [the engine] has been disassembled, cleaned, inspected, repaired as necessary, and reassembled; and (2) It has been tested in accordance with approved standards and technical data . . . acceptable to the Administrator . . ."

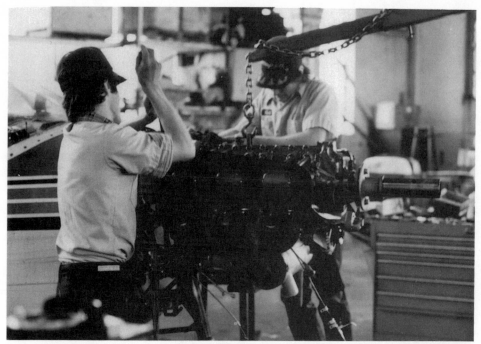

Fig. 10-1. *Mechanics at an overhaul facility prepare a completed engine for reinstallation into the aircraft.* Photo courtesy of T. W. Smith Engine Co., Inc.

This very vague wording requires that an overhauled engine be restored to manufacturer's approved "service limits." While there are FAA guidelines dictating service limits, it is acceptable to replace an out-of-tolerance part with a used one that is within tolerance. However, the used part might be on the borderline of the tolerance and exceed it after only a few hours of use! From a practical standpoint, few, if any, overhaul shops are going to employ parts that close to tolerances because they would end up replacing them again during engine warranty. After the overhaul is complete, the engine logbook's total time continues, and it is noted that the engine has zero hours since major overhaul (SMOH). See FIG. 10-1.

While the manufacturer would prefer to sell everyone a new or rebuilt engine, that does not meet the needs of all customers. Therefore, most manufacturers also overhaul engines, but unlike the typical non-manufacturer overhaul facility, they will replace all critical parts with new OEM (original equipment manufacturer) parts as a matter of routine. In addition, the OEM uses the latest modifications; independent shops may have an inventory of old, unmodified PMA (parts manufacturer approval) parts (see Chapter 2). Some mechanics also believe that the OEM will do a better job on their own engines than an outside shop would. Lycoming, for instance, will never weld a crack in a crankcase or a cylinder, a common practice at many overhaul shops. Instead, they routinely replace all cylinders with new ones during their overhauls. See FIG. 10-2.

Fig. 10-2. *An engine awaits an overhaul.* Photo courtesy of T. W. Smith Engine Co., Inc.

What's the big deal about welding cracks in a crankcase or cylinder? According to Avco Lycoming's Williamsport Division Metallurgical Laboratory, welding aluminum can lead to significant problems. First of all, it is important to know exactly which aluminum alloy is being used, but that is often proprietary information. A welder might not know for sure, and using an improper welding technique leads to premature failure. Yet another reason is that welding aluminum is not easy; few welders can successfully handle the job, and even the best can leave traces of tungsten, making it an unsatisfactory weld.

Complicating the problem still further, every new crankcase is heat-treated for strength and then machined for exact tolerances. Normal welding techniques can cause distortion of the aluminum and will certainly weaken the part. Heat treating the part again will also distort it, causing the mating surfaces to misalign. It is for that reason that welded crankcases typically leak oil. But the most skillful welder under the best circumstances still might not solve the problem. The original crack was almost certainly the result of fatigue or lack of strength in a critical area. Even after welding, porosity and invisible sub-surface cracks can act as stress concentrators and cause premature failure. Dye penetrant inspection will not reveal this problem. Only X-ray produces a suitable inspection, but few operators have the equipment. All too frequently the result is failure in the same area.

Also be wary of the operator that recommends a top overhaul. They should only be done when needed on the diagnosis of a competent A&P mechanic. Too many engines get "topped" for no reason. The average engine should run to TBO if it received a good overhaul to begin with, is properly operated, is well maintained, and has been flown frequently (at least 15 hours per month).

REBUILT ENGINES

The option to overhauling an engine is to purchase a rebuilt one. Rebuilt engines can only be done by the OEM because the finished engine is considered "zero time." That means it is legally a new engine, with a new serial number, new logbook, and new name plate. In fact, it is unlikely you will get your original crankcase back. All tolerances are factory new and only new parts are used in the process. See FIG. 10-3.

CHOOSING AN OVERHAUL FACILITY

Considering the options, where to get the work done can be a difficult question. While the average FBO (fixed-base operator) doesn't have the equipment, expertise, or manpower to perform engine overhauls, that shouldn't stop an owner from contracting with them to have the work done. Your local FBO acts as the liaison between you and the overhaul facility. You give them your airplane with the old engine and they give it back to you ready to go. If anything goes wrong later on, the FBO will be much more accessible than some distant overhaul shop.

As liaison, the FBO does the R&R (removal and replacement), and often overhauls the accessories while the engine is out; the airframe never leaves your home airport. They handle all the engine shipping and paperwork. And because most overhaul shops give FBO's a 10 to 30 percent discount not available to the customer, the cost is the same as if the customer took it to the overhaul shop himself. Finally, you are supporting your local FBO at no cost, and significantly less trouble, to you.

If you prefer to contract directly with an overhaul facility, consider doing the following. There are numerous ads in trade magazines. Contact some of overhaul shops and listen to what they have to say. Then ask for and call a number of references. Discuss your intentions with your local A&P mechanic; find out where they are sending their engines for overhaul. Ask prospective companies for a copy of their warranty and look it over carefully. A warranty says a lot about what the facility thinks of its own work. A very short warranty period might indicate the routine use of used replacement parts close to

Fig. 10-3. *A refurbished crankshaft and connecting rods are ready for the engine's rebuilding.* Photo courtesy of T. W. Smith Engine Co., Inc.

their service limits. Random samples ranged from 100 hours in one case to 6 months regardless of hours in another. Lycoming offers the same warranty for its overhauled engines as it does for the new ones—a strong statement indeed! After the initial warranty period expires, most facilities go to some form of prorated system where the customer pays a greater percentage as the number of hours on the engine increases.

While the warranty is important, so is proximity and immediate service. How available is the facility to solve problems after the overhaul is finished? Most problems tend to turn up in the first 50 hours of operation. Though it is not always possible to find the right facility near-by, location is still an important consideration.

For owners who rely heavily on aircraft availability, there is yet another option: an exchange program. The overhauling shop determines a value for your existing, run-out engine, deducts it from the price of an overhauled one, then you swap engines. Downtime can be as little as a couple of days. While the cost is less than a new or rebuilt engine,

Fig. 10-4. *The mechanic inspects the mating surfaces prior to installing the cylinders.* Photo courtesy of T.W. Smith Engine Co., Inc.

it is typically going to cost more than waiting for your own engine to be overhauled. It's worth noting that exchange programs go on the assumption that your cylinders, case, and accessories are in normal run-out condition. If they find unusual wear after they have had an opportunity to disassemble the engine, the customer must pay the difference. The same thing would happen if you were having your engine overhauled yourself. If you prefer the use of OEM parts but don't quite want to pay the price of an OEM overhaul, some operators offer two levels of service. The lowest priced overhaul uses PMA parts, while an overhaul with OEM parts is only available at a higher price. See FIG. 10-4.

The other major question is FAA certification. A shop does not have to be certified to do engine overhauls, although the work must be done under the supervision of an FAA-certified powerplant mechanic. Many overhaul facility owners say FAA approval for the shop isn't necessary, that it is more of an advertising tool than anything else. One manager confided in me that shop certification was a paperwork nightmare. He said, with respect to the engine overhaul, "We don't do anything different now than we would do with [shop] certification. No matter what, the work is done under the authority of a qualified A&P mechanic." A number of FBO managers who send their own engines to various overhaul facilities echoed the sentiment. They were all more concerned about the integrity of the shop's personnel and the shop's track record than whether or not they had FAA certification for the shop.

As you are researching the possibilities, you will discover that prices vary significantly. A good rule of thumb regarding costs is that the least expensive option is to have an engine overhauled by an independent shop using PMA parts. Using OEM parts would add some to the same base price. Next in line would be to have it overhauled by the original manufacturer. The next most expensive would be to have it replaced with an OEM rebuilt engine, and finally, the most expensive option is to replace it with a new engine. All else being equal, you could expect the actual TBO to increase in the same order.

When preparing to make a final decision among several shops, beware of the lowest bidders; check them out very carefully. The 10-minute TBO, though not common, has happened. One unfortunate owner went to a cut-rate shop only to have them use chrome-faced piston rings with chrome cylinders—a definite no-no. The airplane never made it off the runway after it left the shop. Fortunately, 10-minute TBOs are rare but 200- to 300-hour ones are not. Search for the most cost-effective shop. This is not necessarily the lowest bidder, but select the one that does the job right for a reasonable price.

PART 2
Electrical System

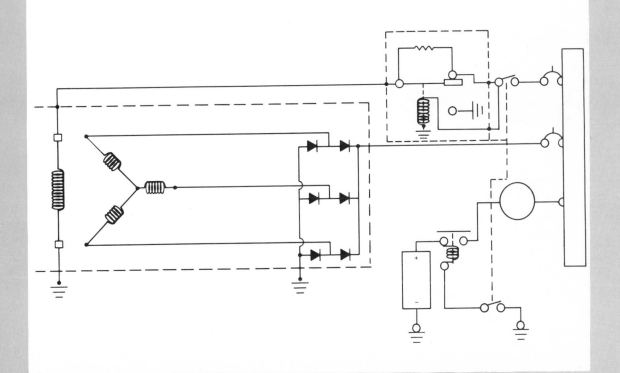

11
Charge Up Your Lead-Acid Battery Savvy

The modern aircraft lead-acid battery is an efficient, carefully designed piece of equipment significantly different from its automotive cousin. It must operate dependably under conditions unheard of for auto batteries. Its list of jobs include engine starting, preflight of electrical equipment and accessories, and backup for the alternator.

The conditions under which it must operate are extreme to say the least. Airplanes go where cars dare not tread: the arctic tundra, high up in the mountains, far into deserts, deep into rain forests, onto rivers and lakes, and into airports below sea level. The battery must operate reliably in unusual attitudes (like inverted flight), at very high altitudes, and it must be capable of handling potential temperature changes in excess of 100 degrees during a single flight (FIG. 11-1).

Despite all its capability, the aircraft battery weighs less and is smaller in size than its automotive counterpart. There are trade-offs though, and the aircraft battery tends to be a little more delicate and does not maintain itself as long, preferring instead to provide a greater short term capacity, such as is needed for a cold weather start. To understand what a battery can do for you and what you must do for it, some theory is necessary.

THEORY OF OPERATION

Three important terms used when discussing the battery are *volts*, *amps*, and *amp-hour*. A volt is a measure of electrical pressure or *potential*. It is the motivating force that moves electrons through a conductor. For example, a 12-volt battery has a potential of 12 volts of electrical pressure. Amps describe the rate of flow or *current*, a measure

Fig. 11-1. *Lead plates within the battery provide a high surface-area-to-volume ratio for compact size and high efficiency.*

of how many electrons flow through a conductor. Amp-hour is a rating given to a battery indicating potential duration of the current flow under ideal conditions. You may think of the whole process as a miniature water wheel (FIG. 11-2).

Referring to the figure, the upper water source and lower water receiver are like the terminals of a battery. As the water flows from one to the other, it expends energy, turning the water wheel. In the case of a battery, it turns a flap motor or illuminates a landing light. The greater the load on the water wheel, the greater the demand for flow to turn it; similarly, the more powerful the electric motor, the greater the demand for current. Eventually, you will exhaust the supply of water in the tank, and so too will you exhaust the battery. A fully charged 20 amp-hour battery in ideal conditions, for example, would be capable of supplying 20 amps for 1 hour, 10 amps for 2 hours, or 5 amps for 4 hours.

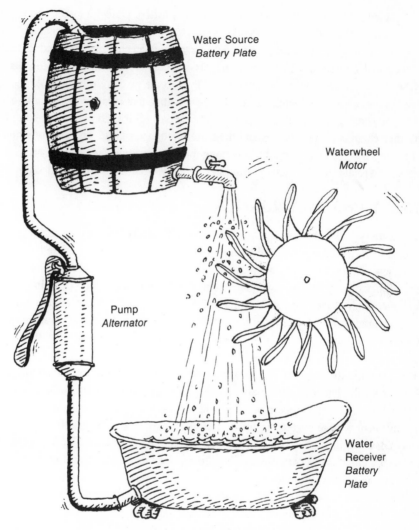

Fig. 11-2. *Analogy of a waterwheel to a lead-acid battery.*

In the water system illustrated, a pump resupplies the source, but in aircraft, an engine-driven alternator recharges the battery continuously whenever the engine is running. It is also sufficient to handle normal inflight power requirements. If the alternator system fails, then there is a limited supply of energy available from the battery.

Murphy's Second Law states: "Nothing is ever as simple as it first seems." Batteries neatly fall under that law. It is a very common misconception that a battery stores electricity; actually it converts electricity into chemical energy during recharging, stores the chemistry, and reconverts it when the battery is connected to some demand such as a hungry starter. Understanding the basic chemical process gives a clue to potential problems. The three primary chemicals involved are lead dioxide in the positive plates, another spongy form of lead in the negative plates, and a liquid electrolyte composed of sulfuric acid and water.

When the battery terminals are connected externally to an electrical demand, the sulfate in the electrolyte combines with the active material in both types of plates and the electrons flow from the negative terminal through the starter to the positive terminal. At the same time, both plates are accumulating a coating of lead sulfate that is highly resistant to current flow. As the coating increases, the battery gets weaker until both plates are completely coated, then chemical action ceases and the battery stops working. The phenomenon that makes the battery practical is that by reversing the current flow, the entire process works in reverse and *re*charges the battery; this is essentially what the alternator does in flight.

BATTERY CHARACTERISTICS

The battery, like the human body, reacts differently to varying environmental conditions. During cold weather, it does not produce as much voltage because the chemical reaction slows down. It is not uncommon for a lead-acid battery to lose as much as 40 percent of its capacity in warm weather (TABLE 11-1). During winter, it is important to change to a lighter grade of oil and change it often to keep it clean. An engine with cold, dirty, heavy grade oil is very difficult to start. Pulling the prop through several times helps break up the oil, but the best course of action is to warm the engine oil with a preheat and perhaps even use a ground power cart for starting.

To determine the state of charge of a battery, it would seem logical to simply put a voltmeter on the battery's terminals and read the open circuit voltage, but remember Murphy's Second Law! Temperature has a very strong effect on cell voltage and there is no convenient way of correcting for it. Determining the specific gravity (sp gr) of the electrolyte is a more reliable method. It is true that temperature also affects sp gr, however it is easily calculated (TABLE 11-2). The hotter the electrolyte, the lower its sp gr.

To test the sp gr of the electrolyte, use a fairly inexpensive and readily obtainable tool called a hydrometer (FIG. 11-3). As the battery produces current, the acid transfers from the electrolyte to the active material in the plates; therefore, less acid remains in the electrolyte. A chemist will tell you that the specific gravity of acid is considerably greater than that of water, so the loss of acid causes a drop in the specific gravity of the electrolyte.

Table 11-1. Percent of Battery Power Available
at Varying Temperatures

Temperature °F	Full Charge	Half Charge
80	100%	46%
32	65%	32%
0	40%	21%

Table 11-2. How Temperature Affects Specific Gravity

°F	Specific Gravity
107	1.260 - 1.280
77	1.280 - 1.290
47	1.290 - 1.300

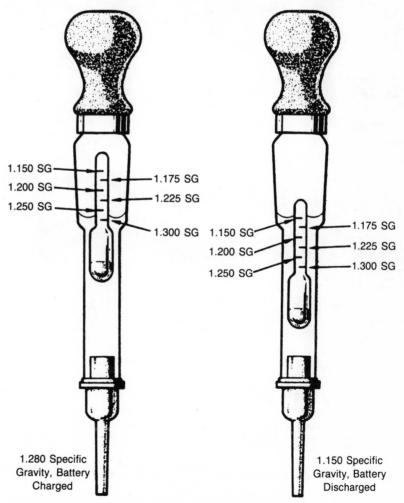

Fig. 11-3. *Typical hydrometer readings.*

To check the sp gr of the electrolyte, take the cap off of a battery cell and draw electrolyte up into the hydrometer. The small numbered stem floats inside the instrument, and the sp gr reading is taken at the fluid level. It is important to note that the stem must be floating for the reading to be accurate (therefore it must have liquid below and around it). A fully charged cell should read between 1.275 and 1.300 (depending on the manufacturer), it is a good idea to check all the cells of the battery. Readings from 1.200 to 1.240 indicate a low state of charge, and engine starting may be difficult, if not impossible. See FIGS. 11-4 through 11-6.

Fig. 11-4. *To check the electrolyte level of an aircraft battery, the protective covering must be removed.*

Fig. 11-5. *After removing the cover, cable contacts should be checked for corrosion and cleaned if necessary.*

If the temperature of the electrolyte is between 70 and 90 degrees Fahrenheit, then the reading is accurate; for temperatures above or below that range, a correction factor must be applied (TABLE 11-3). For instance, if the electrolyte temperature is 100 degrees and the specific gravity reading is 1.267, the chart tells you there is a +.008 correction factor. Therefore, 1.267 + .008 = 1.275 sp gr.

Lombardo's First Law of Reciprocal Reliability states: "The reliability of any person or thing is inversely proportional to the time available to accomplish the task." Practically speaking, that means batteries tend to fail most often when you are in a hurry; therefore, the temptation is great to put the battery on a fast charge. Be careful, because there is

Fig. 11-6. *A hydrometer can be used to test the specific gravity of the electrolyte or to remove solution during maintenance procedures.*

Electrolyte Temp. (°F)	Correction Factor
120	+.016
110	+.012
100	+.008
90	0
80	0
70	0
60	−.008
50	−.012
40	−.016
30	−.020
20	−.024
10	−.028
0	−.032
−10	−.036
−20	−.040
−30	−.044

Table 11-3. Temperature Correction for Specific Gravity Readings

a very strong possibility of overheating the battery and buckling the plates. Unless you are an experienced mechanic, you should limit the charging rate to 4 amps per hour. It is also worth noting that battery charging produces the highly explosive hydrogen gas, so be careful in connecting and disconnecting the charger to prevent possible sparking, and, of course, smoking should never be permitted in the area!

I once had a cat that if left alone for too long would repay my neglect by purposely eating a prized plant or knocking books off a shelf. Batteries are a lot like that—if left alone too long, they tend to cause problems. A battery not used for a month can easily self discharge to 50 percent capacity. Temperature is a significant contributing factor to the rate of self discharge. The higher the temperature, the faster the specific gravity decreases—as much as .003 per day with an ambient temperature of 100 degrees Fahrenheit.

$$100 \text{ °F}: -.003 \text{ sp gr per day}$$
$$80 \text{ °F}: -.002 \text{ sp gr per day}$$
$$50 \text{ °F}: -.0005 \text{ sp gr per day}$$

From the table above, it would appear that the worst time to let a battery sit idle is during the hot months. However, here again, Murphy's Second Law rears its ugly head. The chemical process involved performs a nasty little trick: it causes oxygen to enter the electrolyte and mix with the hydrogen already present. A quick trip to the old high school chemistry book assured me that water is the result. Therefore, a discharging battery keeps adding pure water to the electrolyte, which when exposed to freezing temperatures means a cracked battery. A fully charged battery with a sp gr of 1.285 would freeze at −90 degrees Fahrenheit; a battery with a 1.100 sp gr has a freezing point of only 19 degrees Fahrenheit (TABLE 11-4). The lesson to be learned is whenever you anticipate not flying regularly, remove the battery and store it in a cool place. In addition, approximately every 5 weeks it should be recharged to prevent a lead sulfate build-up on the plates. Lead sulfate (a crystalline formation) is a terrible conductor that can prevent recharging and permanently damage the battery. If despite all your best efforts you discover a dead battery in your airplane, do not jump-start the aircraft. Modern aircraft are equipped with alternators,

Table 11-4. Effect of Specific Gravity on Battery Freezing Temperature

Specific Gravity Corrected to 80°F	Battery Freeze Point (°F)
1.289	−90
1.250	−62
1.200	−16
1.150	5
1.100	19

and while it is possible to activate the starter with a jump, the alternator requires approximately 2 volts of battery power to excite the field coil. In short, no battery, no alternator. In such a situation, you would have absolutely no electrical equipment at all. The engine would continue to run, because the ignition system is totally independent of the electrical system; it is powered by the self-contained magnetos. If, on the other hand, there was very low voltage left in the battery, the alternator would put a very high rate of charge on it and overheating could result.

BATTERY INSTALLATION

When installing a new battery, there are several important considerations, the most important of which is reading the manual supplied by the manufacturer. In all cases, follow the manual! Only a fully charged battery should be put into an airplane, and few new batteries are fully charged. The Gill GSM-682 Service Manual states:

"New batteries are received dry charged and will deliver 75 percent of the rated capacity after the initial filling of electrolyte without further charging, however, it is important that they be given an initial charge to insure their airworthiness before installation in the aircraft."

The biggest problem is that the alternator will charge the battery at a very high rate, which leads to battery overheating and potential buckling of cell plates. Another significant concern when installing a new battery is the mixing of the electrolyte.

I remember a trick I played on my high school chemistry lab partner. I dumped a vial of water into a beaker of acid . . . lots of fun for everyone except the girl downstream and her new angora sweater. She received both a 2nd-degree burn and a sweater modified by Lombardo. I received an "F" in the course. Moral of the story: always pour acid into water, not water into acid! And remember, it really is sulfuric acid; if you spill a little bit on the airplane you will discover that it literally eats through aluminum at an alarming rate. If that does not bother you, then consider that it will cause severe skin burns, blindness, and inhaling the fumes can cause you to permanently lose your sense of smell and taste. If acid comes in contact with your skin, immediately wash the affected area with baking soda to neutralize it. Needless to say, sulfuric acid is very powerful. The simple act of mixing it with water into electrolyte creates heat. Always give the mixture time to cool off before actually pouring it into the battery cells to prevent potential plate damage from overheating.

As a battery charges and discharges over time, the active material from the plates wears off and slowly collects at the bottom of the battery. Like people, batteries get old, and at some point the plates will no longer have enough active material left to produce adequate capacity. When you begin to find that the battery just doesn't hold a charge like it used to, then it is probably time to trade it in for a new one. Due to the tendency for active material to wear off and collect at the bottom, it is important to never attempt to remove the electrolyte by turning the battery upside down and letting it pour out. The sediment will then run over the cell plates and a strong potential for the plates to short out will occur. If, for any reason you need to remove the electrolyte, it should be drawn out with a syringe type instrument or even a hydrometer.

VOLTAGE REGULATOR

It is the job of the voltage regulator to make sure that the electrical system has a constant, regulated source of voltage regardless of the condition of flight. Fundamentally, it is the electrical system's equivalent to a propeller governor. There is a rule of thumb that says if a battery is using too much water, the voltage regulator is set too high; if the battery doesn't seem to stay charged, then the voltage regulator is set too low. Essentially that is true, however, once again Murphy sneaks into the picture. In flight, the battery is recharged by the alternator; this happens because the mechanic has set the voltage regulator to a value slightly above battery voltage. A 12-volt battery system will have an alternator output of 14.25 volts, and a 24-volt battery system will have an alternator output between 28 and 28.5 volts. Because the alternator is putting out a higher voltage, it is in essence force-feeding the battery.

Unfortunately, the voltage regulator makes mistakes, too. If it is set too high, the battery will overcharge, causing excessive operating temperatures. Higher temperatures means the active material will sluff off faster, significantly shortening the life of the battery. Additionally, the hotter temperatures cause the water in the electrolyte to evaporate, necessitating frequent additions of water. Having to add water more frequently than every 50 hours is an indication of high operating temperatures.

In extreme cases, the battery temperature can get so high that the plates buckle or the electrolyte boils and pours out of the battery vent caps. If, however, the voltage regulator is set too low, the battery will never completely charge. This is a tricky area to play detective, because if you have added electrical equipment to your aircraft, it is possible that the alternator is inadequate to meet the demand. The result is the battery has to make up the difference, and instead of being recharged in flight, it would end up discharging by trying to share the load of the alternator. In this case, one set of symptoms could indicate two very different problems: inadequate alternator, and low voltage regulator setting. Playing with voltage regulators can be deceptive, so consult an A&P if you suspect a problem.

PREVENTIVE MAINTENANCE

The best key to a long and happy battery life is good preventive maintenance. Several good rules are:

1. Keep the battery fully charged at all times.

2. Never hit the battery posts with a hammer in an attempt to remove terminal connections.

3. Keep the electrolyte level up by adding water as necessary.

4. When not in use for extended periods, store the battery in a cool, dry area and recharge it at least every 5 weeks.

5. Check terminal connections for corrosion. Corrosion can be removed by gently brushing the surface with a hard bristle brush. Apply petroleum jelly to the studs and terminals to prevent further corrosion.

6. Never pry cable connectors with a screwdriver in an attempt to remove them.

7. Never overtighten terminal bolts, you might unnecessarily jar the battery post.

8. Never reverse battery leads—positive goes to positive, negative to negative.

9. Make sure the battery is securely held down without excessive tightness; if the straps are too tight, you risk damaging the battery container.

TROUBLESHOOTING

Specific battery-related problems for most electrical systems are generally not too difficult for an experienced mechanic to readily solve, provided you do your part to help. Explain the symptoms in detail, noting such things as how long the battery sat idle, what the ambient temperatures were during those days, and the conditions at the time of the problem. It is also helpful if you know what the current draw has been according to the ammeter (if yours is one that only indicates charge/discharge, then report what it indicated); how frequently you need to replace water; whether the battery ever gets fully charged; if you experience difficult starts, even in warm temperatures; what happened when you attempted a start in very cold temperatures without pulling the prop through, getting a preheat, or using a ground power unit; whether there was any failure of specific electrical equipment; the source of any pungent odors that might indicate burning wires; or any unexplainable dimming of lights (TABLE 11-5). Answers to all these questions help the mechanic to diagnose the problem.

In general, the time-tested lead-acid battery is reliable and trustworthy. A bad battery is more often the result of bad preventive maintenance. One of the first things a budding mechanic learns in A&P school is the old saw, "If you take care of your tools, your tools will take care of you!" That saying applies equally well to the lead-acid battery.

Table 11-5. Troubleshooting Batteries

Trouble	Cause	Remedy
Battery will not hold its charge:	Battery life is beyond warranty.	Replace
	Charging rate set too low.	Check and correct the setting in accordance with instructions applying to regulating equipment.
	Discharge too great to replace.	Check battery for proper size and capacity. If too small or too low-rated capacity, replace with proper battery. Use of starter on ground and other electrical equipment in air must be reduced.
	Standing too long (hot climate).	Remove battery and recharge.
	Equipment left on accidentally.	Remove battery and recharge.
	Short circuit, or short to ground in wiring.	Check wiring and correct trouble, then recharge.
	Broken cell partition.	This is usually indicated by two or more adjacent cells running down continually. Replace battery.
Battery life is short.	Overcharge.	This causes buckling of plates, shedding of active material, oxidation of grids, overheating, excessive loss of water. Check and correct adjustment in accordance with instructions applying to regulator equipment.
	Level of electrolyte is below top of plates.	Keep electrolyte level above cell separators.
	Frequent discharges. This is due to excessive use of starter and other electrical equipment while on ground and recharging in air.	Reduce unnecessary use of starter and other electrical equipment while on the ground.
	Sulphated plates. This occurs when the battery is left in a discharged or uncharged (one half or less) condition for a period of time, or electrolyte is not maintained at its proper level.	Charge at normal rate until the specific gravity does not rise for two hours and then give a 60 hour overcharge at 10% of the normal charging rate of the battery. If battery capacity is still low, replace battery.
Cracked cell jars:	Hold down loose.	Replace with fully charged battery.
	Frozen battery due to adding water in cold weather without charging the battery sufficiently afterward to thoroughly mix the water with electrolyte before letting it stand, or due to low specific gravity of the electrolyte caused by improper filling.	Replace with fully charged battery.

Table 11-5. (cont'd)

Compound on top of battery melts:	Charging rate too high.	Check and correct setting in accordance with instructions applying to regulating equipment.
	Electrolyte on top of cells. Caused by overfilling. May short circuit the battery. The resulting heat will then soften compound.	Remove any electrolyte from top of battery and neutralize with solution of sodium bicarbonate or ammonia. Then, wash battery thoroughly. charge and replace in airplane.
Electrolyte runs out of vent plugs:	Too much water added to battery.	Remove excess to correct electrolyte level.
	Excessive charging rate.	Check and correct setting in accordance with instructions applying to regulating equipment.
Cell connector melted in center:	Shorted or grounded cable, causing direct full discharge of battery.	Repair short or ground and replace battery.
Battery freezes:	Discharged battery.	Replace with fully charged battery.
	Water added and battery not charged immediately.	
Polarity reversed:	Battery connected backward on airplane.	Slowly discharge completely and then charge correctly and test before use.
	Battery connected backward on charger.	
Battery consumes excessive water:	Charging rate too high.	Check and correct setting in accordance with instructions applying to regulating equipment.
	Electrolyte runs out of vent plugs.	Level of electrolyte too high. Adjust level.
Battery will not come up to charge:	Battery worn out.	Give capacity test and replace if capacity is too low.
	Plates badly sulphated.	Charge as for sulphated.
	Improper storage. Dry batteries stored in a damp location, or wet batteries stored for too long a period without charging.	Charge as for sulphated plates.

12

Quirks and Qualities of NiCd Batteries

The nickel-cadmium (NiCd), often called "nicad" (pronounced "nye-kad") battery works under the same principle as its lead-acid counterpart (see Chapter 11). Both are electrochemical systems that store and supply electrical energy as needed.

The main components of the NiCd battery are the positive and negative plates, separators, electrolyte, the cell container, and the cell vent. The individual cells (typically numbering 19 or 20, depending on application and manufacturer) contain thin, porous, sintered nickel plates. Polarity is achieved by impregnating the plates with nickel-hydroxide for positive polarity and cadmium-hydroxide for negative.

Opposing plates are separated from one another by a continuous strip of porous plastic that serves as a reservoir for the electrolyte and a sheet of cellophane that acts as an insulator. This sandwich structure is fitted into a case that is typically made of a plastic-like substance called plyamide. The negative plates are then connected to a negative cell terminal and the positive to the positive cell terminal. The two terminal openings are sealed to the case with an external O-ring to prevent electrolyte from leaking out of the case. The electrolyte is a 30 percent solution (by weight) of potassium hydroxide (KOH) in distilled water.

Each cell has a filler cap that is used to service the electrolyte. The cap serves as a one-way valve to vent gas resulting from accidental overcharging, yet it prevents air from entering the cell. See FIG. 12-1.

Individual cells are connected in series along highly conductive bars, often made of nickel-plated copper. The current can flow in only one direction. Because each cell averages 1.2 volts, a 20-cell battery would equal 24 volts. The entire assembly is put into a battery case that is typically made of stainless steel, plastic-coated sheet steel, or painted sheet

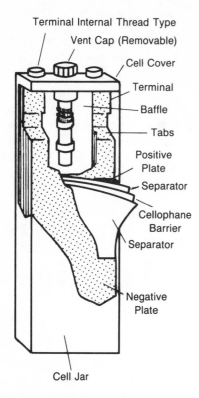

Terminal Internal Thread Type
Vent Cap (Removable)
Cell Cover
Terminal
Baffle
Tabs
Positive Plate
Separator
Cellophane Barrier
Separator
Negative Plate
Cell Jar

Fig. 12-1. *NiCd vented cell.*

steel (FIG. 12-2). The case is attached carefully to a ventilation system to prevent gas buildup as a result of inadvertent overcharging and to permit forced-air cooling during operation. The two end cells are connected, usually by nickel-plated copper links, to the positive and negative battery terminals.

During the discharge process, some electrolyte is drawn into the plate sets, causing the fluid level to decrease. Recharging the battery forces the electrolyte back out of the plates, causing the fluid level to rise to its normal level. The pilot should *not* add electrolyte to a NiCd battery just because the fluid level appears low. Recharging during subsequent operation could force the highly caustic electrolyte to overflow and damage the battery and surrounding area, and a blocked vent cap could result in a cell explosion.

BATTERY RATING SYSTEM

NiCd batteries have a conventional rating system: first the nominal discharge voltage is given—1.2 volts per cell—and then the battery capacity. The latter is the amount of power a fully charged battery provides until a fully discharged condition occurs and is measured in amps per hour (Ah). With the NiCd battery, however, it is important to know the duration (in hours) of the discharge. A nickel-cadmium battery is considered discharged when cell voltage decreases to 1.0 volts. For instance, 40 Ah battery at a 5-hour rate implies the battery would provide a total of 40 amps over a discharge period of 5 hours; that is 8 amps per hour. The catch is it doesn't mean the battery will provide 40 amps for 1 hour or 20 amps for 2 hours. The Saft operating and maintenance manual says the higher the current drain, the less the battery capacity. The same battery at the 1-hour rate would have only a 34 Ah capacity; it would supply a 34-amp current flow for 1 hour before the cell average dropped to 1.0 volts.

Under equal conditions, NiCd batteries long outlive lead-acid batteries. They provide faster starts, recharge more quickly, and maintain a higher state of charge longer. While the optimum temperature range is listed at 60 to 90 degrees Fahrenheit with recharge cautions below −20 degrees and above +120 degrees Fahrenheit, NiCd batteries operate more efficiently at temperature extremes than do lead-acid batteries. Many operators used NiCd batteries routinely at temperatures ranging from −30 through +130 degrees Fahrenheit. Other advantages are reduced engine wear as a result of faster starts, low internal battery resistance, and an inherent ability to maintain high power output longer than an equivalent lead-acid battery.

Cost alone is a significant deterrent to owning a NiCd battery; they are often four to five times as expensive as a lead-acid battery. Servicing is more critical and the battery requires constant temperature monitoring when in use. The worst drawback, though not a common problem, is called *thermal runaway*.

THERMAL RUNAWAY

More properly called *overcharge runaway*, thermal runaway results in self-destruction of the battery. The causative factors are heat, reduced resistance, and high current flow. An overcharge runaway scenario goes something like this: the aircraft has been flying short trips all day in instrument conditions. There have been frequent battery-powered engine starts, and the electrical load was continuously heavy during start and inflight. In other words, the battery has been getting a heck of a workout. This causes a constant, excessive charging of the battery that generates excessive battery temperature.

These factors, combined with as few as one bad plate in a cell, present the beginning of the NiCd battery's equivalent of core meltdown.

When plates short, they overheat, which causes the entire cell to overheat. It, in turn, overheats the surrounding cells, and as internal battery heat increases, it eventually will begin to decrease the cell's internal resistance. Lower resistance permits higher current flow. The more current that flows, the more heat that is produced and the lower the resistance is. Before long, the cellophane between the plates in the bad cell deteriorates, making the situation worse. With even less resistance, more current flows and this begins a vicious cycle of increasing temperature and decreasing resistance. Thus far, it is the generator that is feeding the battery; if the problem is caught in time, merely isolate the battery from the generator and the problem should end. But there is an even more serious, possible outcome.

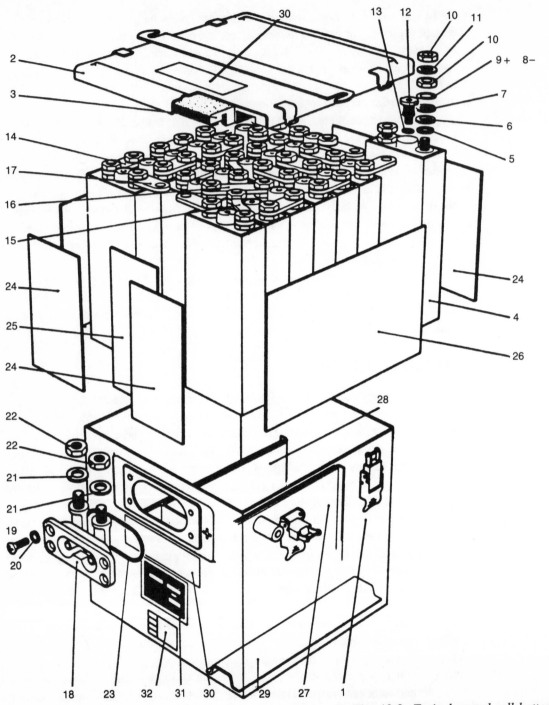

Fig. 12-2. *Typical vented cell battery.*

Table 12-2. Key to Diagram

Figure No.	Nomenclature	Quantity Per Battery
1	• Complete container	1
2	• Cover	1
3	• Cover Gasket	1
4	• Cell	20
5	O-Ring	40
6	Washer	40
7	Spring Washer	40
8	Negative Polarity Washer	20
9	Positive Polarity Washer	20
10	Nut	80
11	Spring Washer	40
12	Vent Plug	20
13	O-Ring	20
14	• Connector	15
15	• Connector	3
16	• Connector	1
17	• Connector	2
18	• Battery Socket ELCON Type BR 8-1	1
19	Screw	4
20	Concave Lock Washer	4
21	Spring Washer	2
22	Nut	2
23	• O-Ring	1
24	• End Wedge	5
25	• End Wedge	1
26	• Side Wedge	2
27	• Insulating Spacer	2
28	• Insulating Spacer	2
29	• Insulating Strut	3
30	• Instruction Plate	2
31	• Identification Plate	1
32	• Modification Record Plate	1

At some point, the resistance of the bad cell will be low enough that the good cells will have sufficient power to feed it. When that happens, isolating the battery does not remedy the situation, because the battery is feeding itself—self-destructing! At this point, the pilot can only hope to land the airplane as quickly as possible. Such a condition has been known to cause a fire in the battery box, and in at least one case, the battery actually has melted through the box and dropped out of the bottom of the aircraft. Again, it is not a common event, and good operational and preventive maintenance procedures are strong deterrents. To aid the pilot in heading off such a problem, a monitoring system is installed with NiCd battery systems. Depending on the manufacturer, it measures either rate of battery charge (a good indication of battery heat buildup) or actual battery temperature.

PREVENTIVE MAINTENANCE

Most manufacturers consider it good preventive maintenance to check new batteries every 50 hours for the first few months of operation. After a while, a pattern will emerge, and the time between further checks can be adjusted accordingly. Whenever working with a NiCd battery, it is extremely important to remove jewelry and any metal articles from your body. Touching points of opposite polarity can literally weld the article to the battery and cause injury.

Electrolyte is extremely caustic and should never be left on aircraft skin or in the battery case. Always rinse and clean the area thoroughly after spillage. Obviously, it will also burn clothing and skin very quickly. Always wear protective clothing when servicing a NiCd battery. This should include at least rubber gloves, rubber apron and protective goggles. If electrolyte should ever get into your eyes, quickly flush them with a large amount of water and get medical attention immediately. Skin burns should be treated by flushing the area quickly with water, then neutralizing the chemical with water and a 3 percent solution of acetic acid, vinegar, or lemon juice. A 10 percent solution of boric acid also may be used. Remember to be very cautious about any battery's fumes; work only in a well-ventilated area. An overcharged battery gives off hydrogen and oxygen gases which together can be explosive. For that reason, make sure battery terminals are tight to prevent sparking.

The hydrometer, a useful tool in determining the state of charge in the lead-acid battery, is useless in testing the NiCd battery. NiCd electrolyte shows no density or composition change during the charge/discharge process, so it is impossible to use electrolyte specific gravity, fluid level, or even battery voltage to determine the charge of the battery. The only reliable way to determine a state of charge is to discharge the battery at a constant rate, according to the manufacturer's instructions, accurately timing how long it takes. An experienced mechanic can determine state of charge based on that information.

Whenever the battery is removed from the aircraft, it should be discharged according to the procedure outlined in the manufacturer's manual and then serviced. Carefully check for general condition. A light powder deposit of potassium carbonate might be noticed on the top of the cells; clean off the deposits. Do not use petroleum spirits, trichloroethylene or any other type of solvent, as they are harmful to the battery. Brushes also are not a good idea, because they tend to force particles between the cells. Use a clean cloth to wipe both the battery case and cover. If there is particulate matter inside the case, use filtered, compressed air to blow it out, and wear safety goggles! During the visual inspection, check for cell-vent integrity. There should be no obstructions or damage to the vent.

When inspecting the battery, if the cell electrolyte level is low, you can add either distilled or demineralized water. Never use tap water, because the minerals will cause shorting of the plates. Never add water unless the battery is fully charged, because the electrolyte level lowers as it is absorbed during discharge. When the battery is brought up to full charge, the level will raise. Approximately three hours after a battery has been charged, the electrolyte level should be ⅛-inch above the visible insert in the cell. If not, fill it to that level.

If a battery is partially discharged when removed from the aircraft, charge it. If it has been inactive for more than two weeks but less than two months, it should be charged before attempting to use it again. A NiCd battery that has been inactive for more than two months should be reconditioned. This consists of completely discharging and recharging the battery according to the manufacturer's recommended procedure. Often referred to as "deep cycle," it is a preventive maintenance procedure that ensures all the cells of the battery are pulling together. In addition, most problems relating to the battery's state of charge are solved by this reconditioning process.

It is difficult to fix a specific number of flight hours for preventive reconditioning. There are many factors that affect the frequency. According to the Operating and Service Manual published by General Electric for its NiCd battery, all of the following are considerations: how well the battery and aircraft electrical system is matched; battery and electrical component maintenance; geographic location and season of the year; operator techniques for engine starting; frequency and severity of engine starts; and battery operating temperatures.

Sometimes a battery does not seem to supply its rated power. If the battery is frequently discharged at a rapid rate then quickly recharged, a cell voltage imbalance can result. The corrective action is to recondition the battery. In the absence of other information, have the battery checked every 50 flight hours until you have sufficient historical data to increase or decrease frequency of servicing. The range can be from a few hours to more than 1000!

Preflighting the NiCd battery is relatively simple. Check the box for structural and chemical damage. Vents and cooling lines should be checked for leaks, obstructions, or any other damage. If there is excessive electrolyte spillage or heavy water use by one or more cells, it could be the result of any one of several problems. The cell could have been overfilled or filled when the battery was in a state of discharge. The vent cap could be loose or there could be damage to the O-ring or vent cap. The voltage regulator charge voltage might be set too high, overcharging the battery. If a visual check doesn't reveal any problems, check the voltage regulator.

TROUBLESHOOTING

Foaming during charge is an indication of low electrolyte concentration. Recondition and replace the electrolyte in the foaming cells. Contamination of the electrolyte can be more serious because the contaminants could have caused permanent damage in the cells. This probably is going to require replacement of the affected individual cell.

If the battery output voltage is below normal, the most likely reason is you have accidentally left a load on the battery and it has discharged. Simply recondition the battery.

The simple physical problems include loose, dirty, damaged, or broken hardware; or loose, corroded, burned, or pitted connectors. Often a thorough cleaning of the problem area can rectify the situation. The worst case is a defective or reversed cell that should be replaced.

If the battery has a low capacity, the problem could be as simple as insufficient charge; just charge it. If the electrolyte level is too low, there is a cell imbalance or the charging voltage has been too low. The battery should be reconditioned and the level of the electrolyte brought up. The voltage regulator also should be adjusted in the case of a low charging voltage. Complete loss of battery output could be the result of broken or disconnected hardware links, or a loose battery connector. Repair or replace the parts as necessary and recondition the battery.

Discolored, corroded, and/or burned hardware, connectors, or terminals indicate improper maintenance. The most likely problems are loose hardware, improper matching of parts, shorts between links, and improper cleaning. Correct the problem and recondition the battery if necessary.

If a cell is distorted or damaged, the problem could be a cell with an internal short, an overheated battery or improper cooling. It also could be the result of charger failure or a plugged cell vent cap. Find the problem, correct it and recondition the battery. Distortion of the battery case and/or cover could be the result of a major explosion of one of the cells, a dry cell, a high charge voltage, a charge failure, or a plugged battery vent. Again, isolate and correct the problem and recondition the battery.

13

Aircraft Alternators: The Current Solution

There are several reasons why virtually all modern aircraft come equipped with alternators rather than the once-popular dc (direct current) generators. An alternator has a higher power-to-weight ratio, and weight is always a consideration in aircraft. More important, an alternator produces more power at a lower engine rpm than a generator does—even at ground idle—a significant benefit for aircraft with a panel full of electronic equipment. Also, an alternator requires less maintenance and servicing than a generator does. The alternator is a simpler system, and because it has self-limiting current, a current regulator is not required.

High on the list of reasons for reduced maintenance is the fundamental design difference between alternators and generators. A generator has brushes and commutators that channel the high current flow out of the rotating armature. This leads to electrical arcing, commutator—bar burning, and rapid brush wear, all of which lead to high maintenance costs. Also, the rotating part of the generator is heavy, which leads to greater wear on the bearings. The alternator does not have these problems, because it is connected to the external circuit by slip rings instead of a commutator, and the armature (called a stator) is a stationary member. The electromagnetic field (EMF) becomes the rotating member (rotor) and turns within the stator. In an alternator, the high current of the stator can go through a set of fixed leads rather than through brushes and a rotating commutator, as it does in a generator. It is no surprise that by 1964, most light aircraft manufacturers had replaced the standard generator with the alternator.

In all fairness, the alternator has some problems, too. Alternating current (ac), because of its expanding and collapsing nature, causes ''noise'' in avionics. The solid-state alternator is, in general, more prone to electrical damage than the generator, which has mechanical relays; improper polarity can literally destroy it. And, an otherwise healthy regulator can burn out as a result of an unrelated alternator problem. From an operational standpoint, there is one significant drawback to the alternator; it requires approximately 2 amps of electricity provided by the battery for it to work. Once the alternator begins to produce current, it becomes self-exciting and will continue to run, despite the condition of the battery. But, if the battery is dead before engine start, you're in trouble. Hand-propping the aircraft may start the engine, but the alternator will never produce current without at least 2 amps from the battery.

A light-aircraft alternator is virtually identical to its automotive counterpart. The only real difference is the aircraft alternator has a holder and special brushes for high-altitude operation. The rotor is a single-field coil encased between a pair of four-poled iron sections on a shaft with insulated slip rings at one end and a nut and washer at the other (FIG. 13-1). This is connected to a drive pulley that runs off the engine accessory section (or crankshaft).

Unlike dc shunt generators, which contain a permanent magnet and have residual magnetism, the alternator rotor has little or no residual magnetism to form the necessary lines of flux to start current flow. The alternator rotor requires approximately 2 amps to make the alternator self-excite and produce current; it is something like using water to prime a dry pump. Before starting the engine, this "priming" voltage is supplied by the battery (FIG. 13-2). Once the engine is running and the alternator is producing current, it supplies its own 2-amp excitation.

Typically, the alternator field switch is interlocked with the battery master switch, which allows the pilot to shut down the alternator without turning off the battery. When the battery master is turned on, the bus bar is energized, sending power through the alternator field switch, then through brushes and slip rings to the rotor, and finally to ground. The primary difference between the brushes on an alternator and those on a generator is the alternator brushes handle only a few amps, whereas the generator brushes handle up to 60 amps.

As the rotor turns inside the stator, its magnetic field cuts through the stator's three single-phase windings that are spaced so that the voltage induced in each is 120 degrees out of phase with the voltages in the other two windings. This Y-shaped arrangement is what produces the three-phase ac, and system voltage builds up rapidly.

Most aircraft systems use dc voltage, so the alternator's ac output is converted (rectified) to dc through an integral silicon diode rectifier. The diodes act as one-way doors, with very high resistance to current flow in one direction and low resistance in the other. They only permit current flow from the alternator to the battery. The diode prevents flow reversal, inherent in ac, thereby rectifying it to dc. The dc voltage resupplies the battery and is directed to the aircraft's main bus through a 60-amp circuit breaker where it supplies the electrical system. It also serves as the rotor's exciter current.

As the demand on the alternator increases or decreases, voltage is varied by a regulating field current. Dc output is fed to a regulator-voltage sensing coil via the 2-amp circuit-breaker alternator field switch. The coil works like a governor, sensing electric bus volt-

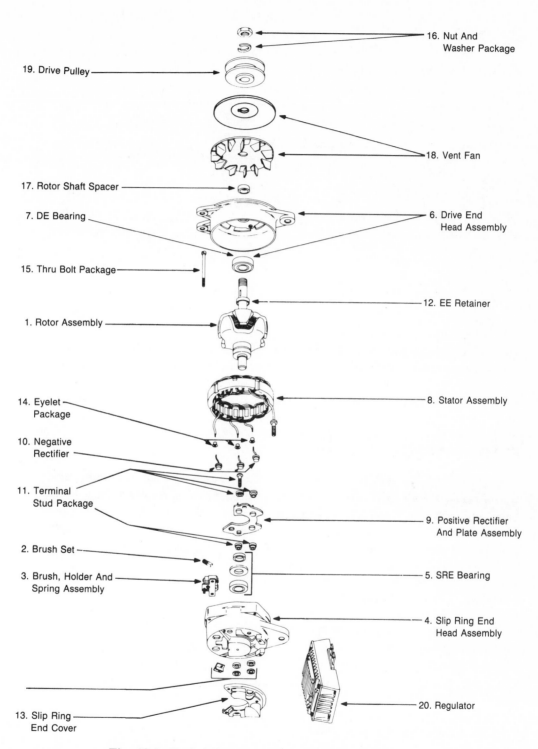

Fig. 13-1. *Exploded diagram of an aircraft alternator.*

16. Nut And Washer Package

19. Drive Pulley

18. Vent Fan

17. Rotor Shaft Spacer

7. DE Bearing

6. Drive End Head Assembly

15. Thru Bolt Package

12. EE Retainer

1. Rotor Assembly

14. Eyelet Package

8. Stator Assembly

10. Negative Rectifier

11. Terminal Stud Package

9. Positive Rectifier And Plate Assembly

2. Brush Set

3. Brush, Holder And Spring Assembly

5. SRE Bearing

4. Slip Ring End Head Assembly

20. Regulator

13. Slip Ring End Cover

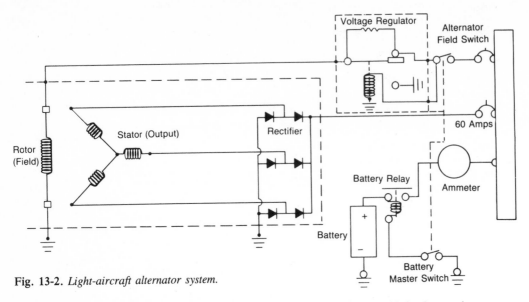

Fig. 13-2. *Light-aircraft alternator system.*

age and varying system resistance as necessary. If the bus voltage is too high, the sensing coil shifts the position of a movable contact which puts a resistor in the circuit. This reduces the field excitation, thereby reducing the alternator output voltage. If the bus voltage is too low, the sensing-coil spring pulls the contacts back and removes the resistor from the circuit, increasing field excitation, thereby increasing alternator output voltage.

SOLID STATE REGULATORS

Newer alternators, such as those made by Prestolite, use solid-state regulators. This type of regulator replaces the sensing-coil mechanism and moving resistor contacts with transistors and a zener diode. The transistor acts as an electric current on/off switch and has no moving parts. The zener diode allows current to move in only one direction, except at a specified voltage value, and then it reverses direction. This serves as a voltage-sensing device to vary current in conjunction with the transistors.

OVERVOLTAGE PROTECTION

Some aircraft manufacturers incorporate an overvoltage control system. This protects the charging circuit from malfunction with a mechanical relay and solid-state triggering device. If the voltage reaches a preset value, the relay opens and the alternator field circuit disconnects. The relay remains open until the alternator switch is turned off. The electrical system then is supported entirely by the battery, unless a backup alternator is installed.

PREFLIGHT INSPECTION

If practical, a preflight inspection should include a good look at the alternator. Mounting bolts should be tight and clean, because they form an electrical connection to the aircraft. In fact, they should be tightened to specific torque value: too loose and the unit vibrates and shifts in its mounting; too tight and the lugs and/or brackets might crack or break.

The drive belt, which also must be properly torqued, should not be so loose as to allow slippage and loss of alternator output. On the other hand, if it is too tight, the belt can break. Even worse, it can cause a side load to occur on the alternator shaft, imposing an abnormal load on the bearings and seals. Such a condition leads to an early failure of the unit. If you can see the alternator fan, visually inspect it for general condition. It should not have any cracks (something to watch for particularly in the area of the welds). The fan should not appear bent and it should have sufficient clearance to turn without scraping a baffle or other structure.

One of the most common complaints about dual alternator systems is an imbalanced output reading. According to Rick J. Clay, technical services representative for Prestolite, if both alternators use a single regulator, it's not uncommon for an imbalanced output reading to occur. Dual generator systems needed a balanced output to prevent component damage, so many pilots became accustomed to watching for an imbalanced output reading.

Output readings on modern dual alternator systems do not have to be balanced. Variations in resistance characteristics between two charging circuits cause imbalanced indications. This could be the result of voltage drops in the charging system, wiring, or ground-circuit connections or even variation in manufacturers' tolerances between alternators. For instance, if the load requirement from two systems were 60 amps, the voltage regulator would allow sufficient current through the field circuits to meet that requirement. The charging circuit with the lesser amount of resistance will use more current than the other, but between the two, they will produce the necessary 60 amps.

If the output fluctuates or there is no output at all, the most likely cause is a loose or broken alternator belt. If you recently have added an alternator or made some related system change and you find that either the aircraft lights flicker, there is poor alternator voltage regulation, the instruments oscillate, or you hear radio interference, you could have an excessive voltage drop. The most probable causes are improper conductor diameter or length. Aircraft with alternators are more prone to have radio interference because of the constant change of voltage and current signals conducted along and radiating from conventional wiring. These fluctuations are picked up by the avionics equipment, either through their power source or antenna wires. The static almost always can be reduced dramatically, if not eliminated completely, by the strategic placement of a capacitor and installation of shielded conductors between the regulator and the alternator.

LOADMETER vs. AMMETER

There are two basic types of alternator indicator. The zero-left, loadmeter type is used to show how much demand is being placed on the alternator. This does not reflect the charge state of the battery. On the other hand, the zero-center type of ammeter does reflect the charge state, because it is in the battery-charging circuit.

The ammeter is one of the most helpful diagnostic tools. Pilots should become familiar with the type of ammeter in their aircraft and how it reacts to battery discharge, alternator failure and loads that exceed alternator capacity. The alternator will provide very different information, depending on where it is placed in the electrical system. One thing is for sure—current flow means heat, enough heat to cause a fire. The pilot should be aware of trends in ammeter readings to head off a potential problem.

An excessive system-voltage reading on the voltmeter probably indicates regulator trouble. This can lead to several serious problems. The battery might be overcharged, causing severe internal damage. Acid might have been forced out of the battery and damage surrounding equipment; and electrical over-voltage may damage other equipment, including light bulbs. For additional information on loadmeters and ammeters, see Chapter 15.

PREVENTIVE MAINTENANCE

The most significant preventive maintenance for your alternator is to assure proper polarity at all times. If you change batteries, use an external battery boost, or during a fast charge, always assure proper polarity. A mistake here can cause permanent damage to the alternator, its rectifier, and many other electrical system components. For the same reason, be very careful not to reverse regulator leads. It also is important to guard against transient voltages that damage semiconductors in the alternator, radios, and other electronics. A 140-volt transient voltage for more than .001 second will fry a voltage regulator! To help preclude this possibility, always keep the battery in the circuit; it serves as a sort of electrical shock absorber.

Old-style generators required "polarizing" when they were new to assure proper polarity. This was done by attaching a temporary external connection to the field circuit. Alternators *never* require this—it will destroy the semiconductors! Obvious preventive maintenance items include a snug alternator belt and battery in good condition. The voltmeter and ammeter should be checked periodically to assure proper operation. The ammeter alone is one of the most important diagnostic tools of the alternator. Another frequently overlooked but very important preventive maintenance item is the general condition of the cowling baffles and airflow path. Alternator temperature limits are critical and are determined by the type of winding insulation, bearing lubricant, and type of semiconductors used. Even the voltage regulator and overvoltage control have temperature limitations. Baffles naturally take a beating as mechanics remove and replace the cowling. As baffles bend, deform, and crack, the airflow becomes altered and this ultimately leads to premature failure of the alternator as well as other parts.

14

Crank Up Your Starter Savvy

A smoothly running engine is about as close to perpetual motion as you can get. It automatically meters the correct amount of fuel, times the ignition, intake, and exhaust valves, and produces sufficient electrical energy to continue running. One thing it cannot do is start itself. Some force must rotate the engine to make the valves, fuel flow, ignition, and magnetos begin operation.

Airplane engines built many years ago were started by briskly pulling the propeller through its normal arc ("propping"), causing the engine to roar to life. Propping the engine was undesirable not only because there were occasional accidents, but also because it was difficult to turn over the high compression engines.

The starter—sometimes referred to as a cranking motor—is simply a battery-powered motor that rotates the crankshaft fast enough to get the engine started. The basic operating principle is magnetism.

THEORY OF OPERATION

A current-carrying conductor such as a wire tends to move when placed in a magnetic field. In FIG. 14-1, the current is moving away from the reader. This is indicated by an X, which represents the tail of an arrow pointing in the direction of travel. Whenever there is moving electric current, there also is a magnetic field surrounding it. Using accepted electrical theory, it can be determined that in this case, the magnetic field around the conductor is moving counterclockwise.

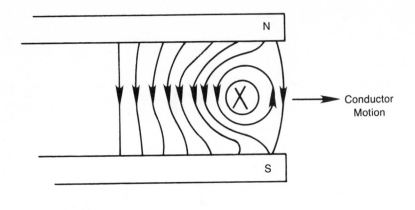

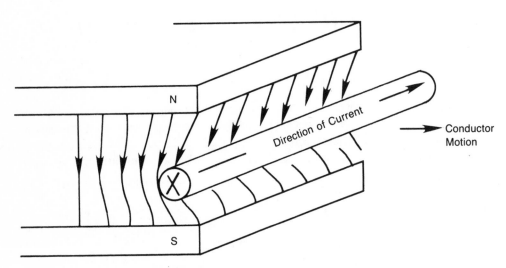

Fig. 14-1. *Interaction of magnetic forces.*

The inherent direction of the magnetic lines in the main field—the north and south poles of the magnet surrounding the wire—move from north to south (top to bottom). The magnetic lines of the two fields on the right side are in opposition and effectively cancel each other. On the left side, the two fields travel in the same direction; the effect is to create a stronger magnetic field. With a strong field on its left and a weak on its right, the conductor moves toward the right.

A single conductor passing through a magnet isn't a very effective tool. However, if we bend the conductor into a ring so it comes back through the magnet, this arrangement can do work. With the current applied in the direction of the arrow, the forces of the magnetic fields tend to push up the left side of the ring, or coil, and push down the right side.

By attaching the coil to a free-wheeling shaft, it will rotate the shaft in a clockwise direction (FIG. 14-2). This primitive, low-power motor can be somewhat enhanced by using heavy copper wire for the coil, which permits heavier current flow and a stronger magnetic field. Similarly, if current is run through windings around the magnetic poles, called *field windings*, there will also be a stronger magnetic field between them. Our single-coil system still would operate poorly, because the torque and speed would vary dramatically as the coil moved from high magnetic influence to low.

The solution is a motor armature composed of many insulated coils connected to an iron or steel core. The core, mounted on a shaft with a bearing at each end, increases and focuses the strength of the magnetic field to maximize its use. As one coil moves out of the magnetic field, another coil enters it. The armature turns continuously and smoothly as long as current is applied.

The starting circuit system includes the starter motor, switch, battery, and load circuit. The battery powers the starter motor and operates electrical equipment whenever the aircraft generator is unable; therefore, a healthy battery is essential to effective starter operation. The starter switch activates and deactivates the starter motor. It completes the circuit between battery and motor. *Load circuit* is the term used for all the cables that connect the individual units of the starter system, including the battery ground strap, the battery-to-starter switch, and the starter-switch-to-motor connection. The load circuit must be of sufficient capacity to carry the necessary starting current with minimal loss to resistance. These cables are chosen carefully to handle high current; a smaller size causes significant reduction in cranking power and can be a fire hazard. Cable connection points should be inspected periodically to assure they are tight and clean, because any unnecessary resistance also reduces cranking power.

The starter motor converts battery electrical potential to mechanical rotary power (see FIG. 14-3). Its frame and field assembly house and support all motor components. The field windings and pole shoes combine with the metal frame assembly to provide a path for the magnetic field in which the armature will turn. A simple but effective brush and holder system feeds the rotating armature with battery power. The brushes seat against, and slide across, the commutator as it turns. Each segment of the copper commutator is insulated from the others and the armature shaft, permitting battery power to go to each individual coil. The armature bearings are seated on the end with the commutator head and on the other end in a pinion housing that also contains the gear drive mechanism that turns the engine.

Various airframe manufacturers employ different methods of activating the starting circuit. Many old aircraft have a T-handle to pull or a solenoid button to push; most modern aircraft have a key switch similar to an automobile. When the switch is turned, battery current goes to the starting motor terminal, which divides it between the field windings around the magnet and the brushes to the armature windings. In both cases, it then goes to ground. The simultaneous current flow through armature and field windings results in a strong magnetic field, which makes the armature turn.

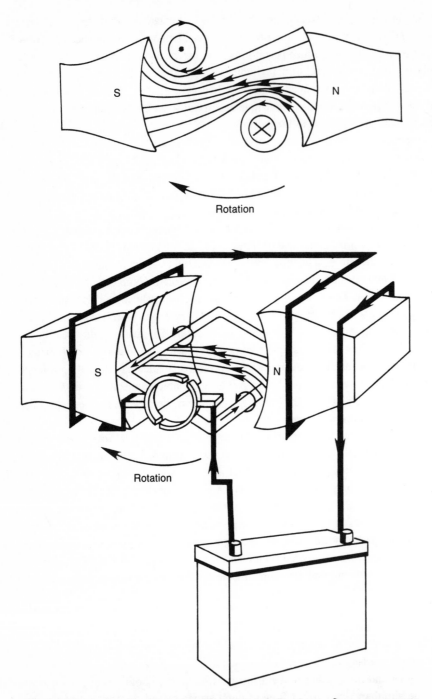

Fig. 14-2. *Interaction of magnetic forces on coil. Dot in top figure represents direction of current flow is toward the reader.*

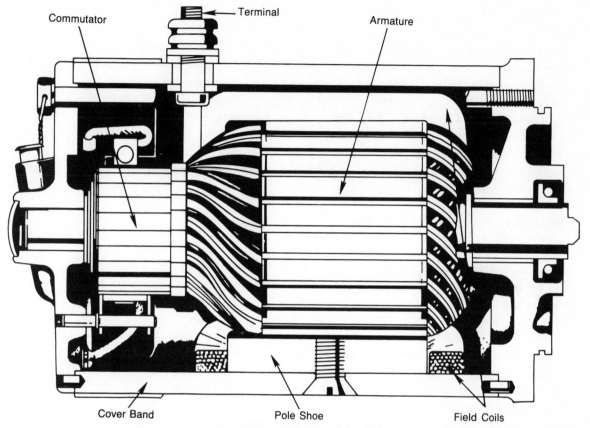

Fig. 14-3. *Aircraft starter motor.*

The armature is connected to a drive system that engages and disengages the motor to the engine flywheel. If you think about how a bicycle works, you can understand why a gear reduction is needed to achieve more torque between the starter motor and the flywheel; without one, the little motor would have difficulty turning the large, high compression engine fast enough to get it started.

STARTER DRIVE MECHANISMS

The two principle types of drive units are the *Bendix drive* and the *overrunning clutch.* The Bendix unit operates by a combination of screw action and inertia. While there is some variation between systems, what essentially happens is the starting motor is engaged, causing the armature and its pinion gear to rotate. The screw action forces the pinion gear forward and it meshes with the engine's flywheel ring gear. After the engine starts, the flywheel begins to turn faster than the armature, so the pinion gear accelerates and is threaded back along a specially designed shaft, out of mesh with the flywheel.

The pinion gear of the overrunning clutch is usually shifted manually, such as with the Cessna 150 T-handle; however, on some models it is controlled by a solenoid. As the gear moves forward, it engages the flywheel and the engine starts; the clutch then releases and the pinion retracts. To assure the pinion does not harm the starter by turning too fast when the engine starts, it is designed to be able to turn faster than the armature without damaging it.

In either type of drive system, if the pinion gear cannot disengage for any reason, the engine might try to drive the starter motor, causing it to burn out. The pilot should never knowingly allow this to happen; engine shutdown is the only solution. Some manufacturers put an annunciator light on the instrument panel to alert the pilot of starter disconnect failure.

PREVENTIVE MAINTENANCE

The best preventive maintenance for the starter system is to use it as little as possible. Another life-extender is to keep the engine in tip-top shape. Problems with the fuel, ignition and lubrication systems result in a harder starting engine, which makes the starter work more than necessary.

Anything that makes it more difficult to start the engine is bad for the starter. For instance, during low ambient temperature starts, use engine preheat and external power. A properly applied preheat helps the crankshaft turn more easily. Another problem associated with cold-weather starts is a possibly weak battery; application of external power assures full amperage to the starter immediately for quick engine cranking.

Running the starter for a long time is especially harmful to the starter motor. Most pilot operating handbooks publish starter time limitations, usually 1 minute or less. The limitation is related to cooling. Cranking the starter for longer than recommended causes the motor to overheat, which can melt the solder that holds the field and armature windings—the starter can literally self-destruct! If the engine does not start within the recommended starter limitation time, something probably is wrong with it. Don't continue to grind away—find out what the problem is with the engine. The three most common reasons an engine won't start are: mixture in cutoff position, fuel selector off, or magnetos off.

Most airframe manufacturers recommend routine starter-system inspection and maintenance at least at the annual inspection. Some handbooks recommend inspection twice a year. The starter is susceptible to cumulative damage and the manufacturer's recommendations should be followed closely. Starter cable terminals should be checked for corrosion; if present, wash the area with baking soda and water solution and dry. Then coat terminals with petroleum jelly. At the same time, check for loose terminal-to-battery stud connection and tighten as necessary. Another vulnerable point is where the cables actually enter the battery terminal. They should fit tightly with no loose strands or partial breaks. Also inspect the general security and integrity of the insulation. These might seem like trivial points, but even loose strands can cause a loss of cranking power.

If there is concern about the system, you can conduct a voltage-loss test to locate high resistance connections that reduce efficiency. Using a low reading voltmeter, have someone crank the engine while you check the voltage loss from the battery post to the start motor terminal. Maximum loss is .3 volts per 100 amps; a 200-amp system would allow a maximum voltage loss of .6 volts. The maximum voltage loss from the battery ground post to the starter frame is .1 volt per 100 amps. Great caution must be exercised when conducting this test. Be sure you disable the engine so it will not inadvertently start; it is best to remove the spark plug leads. Also, stay clear of the prop, as the starter can turn it fast enough for it to be lethal. Have an experienced mechanic do the test, or at least have supervision on how to do it safely.

Because the lead-acid battery is an integral part of the light aircraft starter system, preventive maintenance is important there, too. There are two ways to assure long battery life. First, keep battery fluid at the specified level at all times. Use only pure, distilled water. Anything else will leave mineral deposits and shorten battery life.

The second way to extend the life of a battery is to maintain the appropriate charge level. Take specific gravity readings with a hydrometer every two weeks. You should know what the specific gravity is for your fully charged battery. A reading of .040 or more below the fully charged value is reason to remove the battery and have it charged.

General cleanliness is important to battery life; dirt and corrosion should not be allowed to accumulate. Clean the lead-acid battery with a solution of baking soda and water, but keep it out of the cells. To determine battery condition, an A&P uses high-rate-discharge test equipment, but the operator can get a good approximation using the following method. Crank the engine for 15 seconds with the starter, but keep the mixture control at cutoff to prevent the engine from starting. At the end of the cranking period, quickly measure the battery voltage. A 12-volt system should read not less than 9.6 volts; a lower reading is an indication of a bad battery. For additional information on lead-acid batteries see Chapter 11: Charge Up Your Lead-Acid Battery Savvy. For nickel-cadmium batteries, see Chapter 12: Quirks and Qualities of NiCd Batteries.

TROUBLESHOOTING

The operator can employ several troubleshooting techniques on the starter circuit. If the starter motor power seems insufficient to turn the engine, the most likely problem is the battery. If the battery is sound, inspect the load circuit; perhaps it has a loose connection or a frayed cable. There might also be insufficient lubrication in the engine or something could be binding the engine or propeller. If the starter doesn't activate at all, the first thing to check is the battery and its connections. The starter switch also could be defective; try jiggling it. If the starter draws unusually high current, there is probably an excessive load. Several things could be the culprit. If it's cold, the problem might be congealed oil. Use a preheat and external power for the next start attempt. Also, there could be an obstruction to the propeller, or in the worst case, the engine bearings might have seized.

Occasionally, starters run too fast. This almost certainly is caused by excessive external power output voltage. Should the starter overheat (frequently made obvious by its odor), there could be several causes. Most often it is caused by exceeding recommended starter time limitations. Bad engine bearings can cause the engine to turn with difficulty, which overloads and overheats the starter. Also excessive voltage can overheat the motor. Whenever you smell an electrical problem, it is prudent to stop whatever you are doing and investigate. Similarly, while aircraft tend to be noisy, excessive starter vibration or noise should be investigated. The probable cause is a loose or broken mounting. I can think of few situations less desirable than having a starter motor flying around loose inside an engine cowling during flight.

15

The
Electrical System

Imagine what it was like to be an airmail pilot in late 1918; the freezing cold, windy, wet flights that often terminated in disaster. At best, a pilot could expect to be chilled to the bone; Jennys and De Havillands had open cockpits. During those early days, the sole purpose of electricity in the airplane was to ignite the fuel/air mixture; the exclusive domain of the magneto. As demand increased for electricity to power other equipment, the wind-driven generator was developed.

For many years, lightplanes exclusively used the 14-volt electrical system. Recently, there has been a trend toward the 28-volt system; Cessna adopted it for single engine aircraft in 1979. The fundamentals are the same whether your aircraft has a 14- or 28-volt system. The primary purpose of igniting the fuel/air mixture is still the exclusive and independent domain of the magneto; however, the demand for electrical energy in the airplane has increased tremendously. Even the simplest of aircraft electrical systems power engine starters, cockpit and position lighting, navigation and communication equipment, engine and flight instruments, and accessories ranging from cigarette lighters to inflight telephones. Unfortunately, as the complexity of each such systems increases, the average pilot's understanding of the system decreases. While we can leave the theory of electrons, neutrons, protons, and so on to the engineers, a basic understanding of electrical systems is necessary if a pilot hopes to be able to adequately preflight or determine inflight problems. See TABLE 15-1 for a list of electrical terms.

Table 15-1. Glossary of Electrical Terms.

alternating current (AC)—An electric current which periodically changes direction of flow and constantly changes magnitude.

ammeter—Instrument used to measure current flow.

ampere (amp)—Basic unit of current flow (flow rate); an indicator of the passage of electrons through a conductor. One amp is the amount of current which flows when a force of 1 volt is applied to a circuit with a resistance of 1 ohm.

ampere-hour (amp/Hour)—A rating given to a battery indicating potential duration of the current flow under ideal conditions. It is the quantity of electricity that passes through a circuit if 1 amp has flowed for 1 hour (amps × hours).

bus bar (bus)—Power distribution point, usually a metal strip where several circuits are connected.

circuit—A number of conductors connected together to complete an electrical path.

circuit protection—Devices in a circuit that protect wiring and/or appliances, such as fuses and circuit breakers.

current—The movement of electricity through a conductor.

direct current (DC)—Electric current that always flows in only one direction.

fuse—A metal link which melts when overheated by excess current; used to break an electric circuit whenever the load exceeds a predetermined maximum.

inverter—An accessory that converts direct current to alternating current.

master switch—A pilot actuated switch designed to control all electric power in the aircraft.

open circuit—A break somewhere in a conductor preventing current flow.

parallel circuit—Two or more circuits connected to the same power source and ground.

rectifier—An accessory that converts alternating current to direct current.

relay—An electric switch that is operated by either an electromagnet or a solenoid.

series circuit—A circuit in which the current must flow through all the circuit elements in order, by a single path.

short circuit—Contact between conductors permitting a short, low resistance path back to the power source.

volt—A unit of electromotive force (voltage); a measure of electrical pressure.

voltage regulator—An accessory which maintains a constant level voltage supply despite changes in input voltage or load.

watt—Unit of electrical power, a rate of doing work. In a direct current circuit: watts = voltage times amps.

ELECTRICAL SYSTEM THEORY

Perhaps the easiest way to understand circuitry is to think of it as rivers of electricity. There are two basic types of circuits: *series* and *parallel*. In the series circuit (FIG. 15-1), current flows from the primary bus to the 5-amp circuit breaker and then to the fuel quantity indicator. Therefore, circuit breakers are always placed upstream and in series with the circuit or accessory they are to protect. If the breaker fails, all electric current is cut off from anything "downstream." It is precisely this type of system that evokes the jeers and cheers of my sisters and I each Christmas as we try to find the burned-out bulb in our ancient Christmas tree lights. One burned-out bulb in a series circuit prevents the whole string from lighting. Those under 20 might not be able to relate to that time-honored tradition, because modern tree lights are parallel wired.

In parallel circuits, current flows from the primary bus in the airplane simultaneously

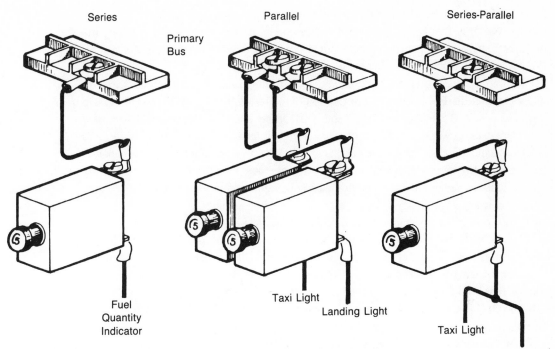

Fig. 15-1. *Electrical circuits showing primary bus and circuit breakers.*

to both 5 amp circuit breakers and then on to both the landing light and fuel quantity indicator. Failure of one circuit, say the landing light, will in no way effect the fuel quantity indicator circuit. This is the ideal situation in aircraft, because all critical electrically driven components would conceivably operate in parallel with each other. Unfortunately, there are so many electrical circuits in modern aircraft that individual protection of each item is not always practical—or necessary.

In the series-parallel circuit, a number of items can be put in parallel with each other (such as the taxi and landing lights) but in series with a single circuit breaker. If one of the lights burns out, the other will continue to work. If however, a short circuit somehow develops in one of the lights and endangers the electrical system, both lights will be isolated from the system when the breaker fails.

SYSTEM SOURCES OF ELECTRICAL POWER

There are three possible sources of electricity that airplanes use: battery, ground power unit, and the alternator or generator. The heart of the electrical system is the battery. Rated in volts and amp-hours, the lead-acid battery is the standard in lightplanes.

It's important to understand that the battery is not an electricity *producer*, rather, it is a *storer*. Theoretically, a 24-volt, 30 amp-hour battery has 24 volts of pressure capable of supplying 30 amps for 1 hour, or 15 amps for 2 hours, or 1 amp for 30 hours, and so on. Unfortunately, the lead-acid battery has many limitations, and the nickel-cadmium battery (NiCd) is finding its way into more and more aircraft. Designed for a long life under extremely adverse conditions, the NiCd has a very low freezing temperature, minimal gas emission during operation, and unlike the lead-acid battery, suffers no deterioration if left in a discharged state. Unfortunately, its high initial cost has hindered its growth into light aircraft. For additional information on batteries, see Chapter 11: Charge Up Your Lead-Acid Battery Savvy and Chapter 12: Quirks and Qualities of NiCd Batteries.

The ground power unit (GPU) is often incorrectly referred to as an APU; the difference is not academic. For the record, the GPU is exactly what its name implies, a ground power unit. It is usually mounted on wheels and is often operated by a flight line attendant. The APU (auxiliary power unit) is an additional small engine, usually a turbojet, whose sole purpose is to supply auxiliary power (electric and often pneumatic) to run aircraft systems on the ground when the engines are shut down, and to aid engine start-up.

Today, virtually all GPU's have the standard NATO 3-pin plug that neatly inserts in a "conveniently" located receptacle on the airplane (like directly behind a propeller or under the wing near the fuselage). Ground power is typically used to start an engine for the first time on a cold day (or other conditions difficult for the battery to operate in). The GPU is also used by the mechanic when electrical power is needed for extended periods of time without the engine running.

When using a GPU, there are several areas of concern. First, make sure all avionics are turned off, because they are highly susceptible to damage from transient high voltage ("spikes"). Most GPUs have a variable voltage output, so it is necessary to make sure the proper voltage has been selected on the power unit to match your aircraft's electrical system. Similarly, the polarity of the GPU must match your aircraft system's polarity. Some aircraft incorporate polarity reversal protection in the aircraft receptacle, however, you should always check it before utilizing the power source. For reasons that are explained later in this chapter, GPUs should not be used to start an aircraft that has a dead battery.

As the demand for a reliable electrical source increased, the wind-driven generator gave way to the engine-driven generator, and though it solved all the pressing problems of the time, it was also marked for obscurity. Large and heavy, the generator had high maintenance costs and long downtime. Another concern was the output variability with engine speed. Geared to the engine, the generator produces electricity at a rate corresponding to engine speed. While it was possible to limit the maximum output of the generator when the engine ran at high speed, there was no way to boost up the low output when the engine was at idle or very low power settings. In a world of increasing electronic sophistication, this became a liability.

For some time, automobiles had been using the alternator that produced ac instead of dc. Not only did it generate more power than the dc generator for a given size and weight, but it was now possible to transmit higher voltage with lower current permitting smaller and lighter weight wiring. Converting ac to dc to run equipment was a simple matter of using a rectifier. By the end of 1964, most newly built planes sported the automotive type alternator (see also Chapter 13: Aircraft Alternators: The Current Solution). Most light single engine aircraft today, such as the Cessna 172RG, use a 60-amp alternator.

According to FAR 23, Airworthiness Standards, for normal, utility, and aerobatic category airplanes, the regulation under which most general variation light single-engine aircraft are constructed, the following requirements exist:

> "Each electrical system must be adequate for the intended use . . . account[ing] for the electrical loads applied to the electrical system in probable combinations and for probable durations."

Elsewhere, the regulation goes on to specify requirements for multiengine aircraft.

> "For each multiengine airplane—there must be at least two independent sources of power (not driven by the same engine . . .''

The electrical system had to be capable of handling anticipated electrical requirements plus be able to keep the battery in a state of constant charge. Another advantage of the ac alternator was its three-phase characteristic, permitting electric motors to shed precious pounds without losing power. Finally, the alternator requires less maintenance costs and downtime.

One of the unique facets of the alternator is its ability to maintain a constant system voltage even during engine idle. This is accomplished by the voltage regulator that monitors system voltage and makes the necessary adjustments to the alternator; if set correctly in a 14-volt system, for instance, it will maintain system voltage ± 0.5 volt.

It might help you to think of the voltage regulator as the electrical system's answer to a constant-speed prop governor. The method employed takes electrical system current and feeds it to the alternator's exciter field in varying strengths. The more current applied, the greater the ac output. Therefore, the alternator requires amperage before it can produce electricity. In fact, the field coil of the alternator requires about 2 amps to make it work. Unfortunately, the alternator needs the 2 amps to begin producing electricity.

Initially, the only source of power is the battery, which is the reason why I earlier stated that GPU's should not be used to start an airplane with a dead battery; in fact, you shouldn't prop-start one either. If the battery is dead, the alternator will not work. With no alternator power, the battery can't recharge, and there will be no power at all to the electrical system. Of course, the engine can continue to run because the ignition system is powered by the totally independent magneto system.

There are several reasons why it is a good idea to have a battery master switch. Some equipment, such as electrically driven gyros, do not have on/off switches; the pilot must be able to isolate them from the battery to prevent draining it when the engine is shut down. Practicality dictates having a convenient method of removing all electrical equipment from the battery simultaneously rather than turning each unit off individually. Safety dictates that during certain situations, such as emergency landings, the pilot should be able to shut down the electrical system to minimize potential fires. Logic says it really isn't a good idea to have wires running through the cockpit carrying potentially high amperage, so the actual master switch is powered by low amperage current. It is this low amperage current that operates an electromagnetic remote relay, located as close as possible to the actual battery, that physically connects and disconnects the battery from the electrical system.

If the alternator becomes inoperative, its exciter field continues to demand system current which is now supplied by the battery. To solve this problem, manufacturers have devised a "split" master. One half of the master switch controls the alternator and the other half the battery. It is possible to remove the alternator from the electrical system and still use the battery. The rule is, if the alternator fails, turn off that half of the master switch.

Those of you who have explorer tendencies might have already noticed a significant difference between the household and aircraft power distribution systems. If you look closely at the cord leading to your table lamp, notice there are actually two wires bound together that connect the lamp to the power source (wall outlet). There is only one wire in the aircraft electrical system. Years ago when aircraft with fabric skin were standard, the electrical systems used two-wire conductors. But the modern all-metal fuselage allows the airframe to act as one of the conductors. The advantages of eliminating half the wiring are obvious: lighter weight, less complexity, lower construction costs, and reduced potential problems. The positive wire runs from the power source to the appliance, and the negative goes from the appliance to the airframe (in some cases, the method of mounting the appliance to the airframe serves as the grounding) (FIG. 15-2).

In choosing the size of the wire to use, there are more considerations than are readily apparent. It must be heavy enough to carry the maximum anticipated load without excessive voltage drop. When replacing wire, the mechanic has to assure that the new wire has the same, or higher, rating. Typically made of copper or aluminum, aircraft wire is designed for adequate flexibility, fire and heat resistance, and in some cases, prevention of radio interference through use of shielding.

All power distribution points (buses) are essentially the same: metal strips to which several parallel circuits are connected. They are typically named for their source of power or the function they serve. For instance, a bus that is directly connected to the battery though the master switch is usually called the battery bus. One that is always connected to the battery with no method of disconnection is a *hot battery bus* or *essential bus* (FIG. 15-2). Items on a hot battery bus will deplete the battery even when the master switch is turned off. Therefore, it is important to assure that all such equipment is individually turned off. While most commonly found on larger aircraft, a limited hot battery bus is installed in some of the lighter aircraft. The Beech 58 Baron uses one to activate baggage and courtesy

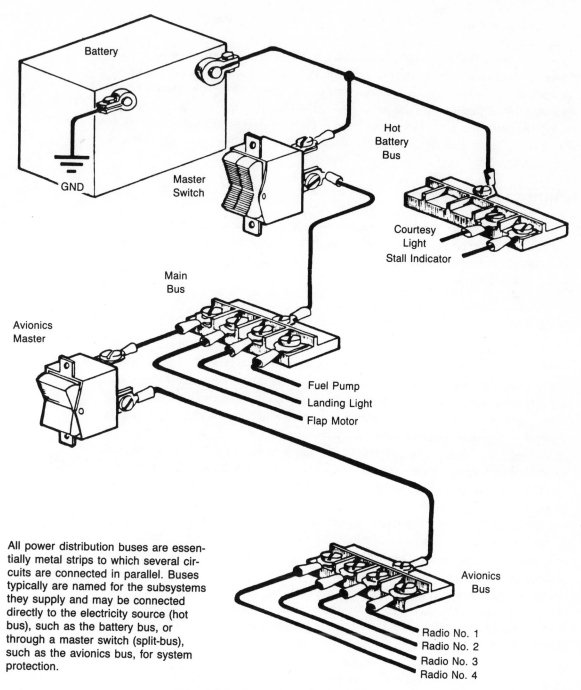

All power distribution buses are essentially metal strips to which several circuits are connected in parallel. Buses typically are named for the subsystems they supply and may be connected directly to the electricity source (hot bus), such as the battery bus, or through a master switch (split-bus), such as the avionics bus, for system protection.

Fig. 15-2. *Power distribution points.*

lights and the stall indicator, where the Cessna 172RG uses it to run the clock and flight hour recorder.

The avionics bus, which is designed as a split bus system to allow separation of the avionics from the rest of the electrical system, is another example of a bus named for its function. The manufacturer typically uses one of two means to remove it from the battery bus: an avionics master switch that must be turned on and off by the pilot, or an automatic relay. The relay method is generally designed so that whenever external power is applied or the starter is activated, the avionics bus is isolated from the airplane's electrical system automatically.

CIRCUIT PROTECTION

When scientists first began experimenting with electricity, there was a need for a weak link in the system to protect both the scientist and the test equipment from being electrocuted. Eventually, two types of circuit protection evolved and are both in modern use: the fuse and the circuit breaker.

Fuses

Initially, somewhere along the circuit early experimenters incorporated an open wire, generally undersized and uninsulated. Because an overload causes an increase in wire temperature, the undersized wire would melt before the rest of the system. This weak, or *fusible*, link was eventually shortened to the name *fuse*. Its primary purpose in aircraft is to protect the system wiring; its secondary purpose is to protect a given appliance. In the modern fuse, higher-than-normal current heats up the fusible element until it reaches its melting point—then the fuse blows. In choosing a replacement fuse, it is important to consider three specifications: current rating, voltage rating, and fusing characteristics.

Current Rating. Current rating is the maximum amount of current a fuse will pass before it melts. What is perhaps not so obvious is the effect ambient temperature has on that rating. Because any given fuse melts at some specific, predetermined temperature, it stands to reason that however that temperature is achieved, the fuse will melt. While it is unlikely that the ambient temperature will ever reach the fuse-melting temperature without the presence of a fire, any temperature increase can increase the rate of failure. The reason for this is that on hot days, the fuse starts out at a higher temperature before the current is ever applied to it. Therefore, at an ambient temperature of 25 degrees Celsius, the fuse ampere rating should be 25 percent higher than the normal operating current of the circuit.

Voltage Rating. Perhaps the most confusing rating of all is the voltage rating. If you look at a fuse, one of the following voltages should be stamped on it: 32, 125, or 250 volts. These ratings indicate the maximum voltage for which the fuse is usable. If the fuse has no rating, you may assume it is 32 volts. The 32-volt fuse is usable for all aircraft dc systems (14- or 28-volt). The 125-volt rating could be used for either system, as long as its voltage level is below that of the fuse. Therefore, fuses should be rated equal to or greater than the voltage of the circuit or equipment.

Characteristics: Slow-Blow vs. Normal-Blow Fuses. The third rating deals with the speed at which a fuse breaks the circuit. There are two basic categories: normal-blow and slow-blow. The normal-blow fuse can be further categorized into fast-acting and medium-acting. These types of fuses are used in circuits where no surges or transient voltages are expected. If a circuit is normally subjected to transient voltage or surges such as engine-starting loads, then the slow-blow fuse should be utilized. Incorporating a built-in time delay, the slow-blow fuse will not immediately fail if the maximum temperature is reached for a short period of time.

The Circuit Breaker

The circuit breaker is another type of circuit protection. In principle, the circuit breaker performs the same task as the replaceable fuse but has the convenience of being able to be reset. In the circuit breaker, the element also reacts to heat, but rather than melting, it expands, causing the circuit to open (trip). Approximately two minutes after it trips, it will have sufficiently cooled down so the pilot can reset it.

If a circuit breaker with a rating over 20 amps trips, this should not be considered a momentary spike and should not be reset. Circuit breakers under 20 amps may be reset as follows: when the circuit breaker trips, the normally flush-mounted core pops out slightly. To reset it, the pilot should first turn off the appliance in question, wait approximately two minutes for cooling, push the core back into its flush position, and turn the appliance on. If the circuit breaker pops again, do not try to reset it a second time. If there are two or more appliances in one circuit breaker, then after resetting it, turn on one appliance and wait. If the circuit breaker doesn't pop, turn it off and try the other appliance to determine which one has caused the problem. If they both work, turn them both on and see what happens.

Occasionally, a circuit breaker pops because of a momentary glitch in the system and there will be no further problem. The only real problem with this type of circuit breaker is because they are flush-mounted, it is impossible to intentionally deactivate a circuit with the breaker—all equipment must be shut off individually. One note of caution: never hold a circuit breaker in if it wants to pop. You would be defeating the purpose of the breaker and this will almost certainly lead to further damage of the circuit and probable fire.

CHANGING FROM DC TO AC

While the aircraft electrical system utilizes dc voltage, some equipment requires 26 or 115 Vac. To accommodate this, an inverter is used to change dc into ac. In light, single and multiengine aircraft, the need for ac is very limited, so an actual inverter is seldom used. Rather, any equipment that requires ac voltage has a type of miniature inverter installed in the instrument itself that greatly reduces the cost and weight associated with installing an actual inverter.

AMMETERS

There are basically two types of ammeters: The *charge/discharge* and the *loadmeter* (FIG. 15-3). The charge/discharge or zero-center type ammeter displays information about current flow. If the needle is to the right of zero, the alternator is working and supplying power to the electrical system. If the needle is to the left of zero, then the battery is discharging, indicating that the alternator is not supplying power to the electrical system. The loadmeter or zero left type of ammeter displays actual current draw (system demand) from the alternator. If the loadmeter reads zero, then the alternator is not supplying power to the system, leaving the battery as the sole source of power.

Whenever alternator failure occurs in flight, all operating electrical equipment begins to deplete the battery. Therefore, the pilot must immediately assess the situation to determine what equipment is absolutely essential to the safety of flight at that moment and turn off everything else to conserve battery power. This procedure, known as *load shedding*, is discussed in the next chapter (Electrical Systems—Load Shedding).

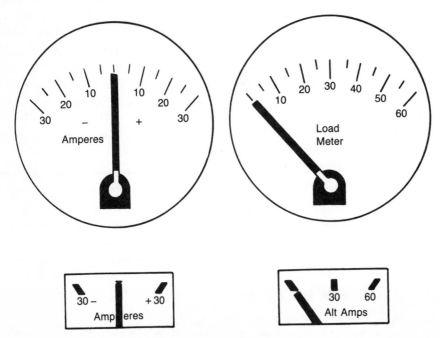

Fig. 15-3. *Types of ammeters.*

PREVENTIVE MAINTENANCE

FAR 43, which outlines maintenance, preventive maintenance, rebuilding, and alteration, only lists three items of preventive maintenance that the owner/operator may perform on the airplane electrical system.

1. Troubleshooting and repairing broken circuits in landing light wiring circuits.
2. Replacing bulbs, reflectors, and lenses of position and landing lights.
3. Replacing batteries and checking fluid level and specific gravity.

Of course there are other things that you can do to help prevent problems. Keep the battery clean, charged, and correctly filled. Always check ground power units for proper voltage and polarity before hooking up to the airplane. Check the security, cleanliness, and condition of wiring that can be seen during preflight. Discipline yourself to include the ammeter in your normal instrument scan, but perhaps the best advice is to generally be aware of the state of the electrical system.

TROUBLESHOOTING

The electrical system is probably one of the most difficult systems in an airplane to understand and troubleshoot. Any information you can give to your mechanic, even seemingly unrelated items, could make a difference. This is primarily due to the fact that many other systems somehow tie into the electrical system. For instance, a hydraulic gear system uses an electric pump. Another quirk of electrical appliances is that some of them continue to operate when there are problems with the system, but their accuracy might be off. Therefore, any tendency of electrical equipment to operate substandard should be brought to the attention of the mechanic. Other occurrences of interest to the mechanic would be unexplainable dimming of lights, smoke or pungent odors that might indicate burning wire, and equipment that operates unpredictably.

One icicled, Illinois day a number of years ago, I attempted to start an airplane and absolutely nothing happened. After several attempts, I gave up and "informed" the mechanic that the battery was completely dead. He wanted to try it himself, but I was so certain about the diagnosis that he pulled the battery and brought it to the shop.

There was nothing wrong with the battery. The mag switch had worn out, probably from the giant key ring full of swinging keys that was always tugging at it during flight. I learned three important things that cold morning: don't hang a bunch of keys from the mag switch; if the battery appears dead, check for excessive key movement in the mag switch; and perhaps the most important of all, let the mechanic do his or her job.

16

Electrical Systems—
Load Shedding

It is a cloudy, wintery night as you fly home after a much-needed weekend getaway with the family. As the airplane momentarily slips out of the clouds, you catch a glimpse of the stars, but the ground remains a mystery beneath the soft, billowy, low-level overcast. Your home airport, still an hour away, is reporting weather that should mean an uneventful instrument approach. The hum of the engine combined with the silent efficiency of your autopilot gives assurance that you are at peace with the skies.

As you scan the glowing instruments, you can see the reflections of your napping family. Suddenly, the sleeping faces are bathed in red light. The alternator light has illuminated; from now on, the only source of electrical power is the battery. You need to reduce electrical load to the bare minimum (called load shedding). What you do in the next few minutes will make the difference between a flight to reminisce about during hangar flying sessions and one that could terminate in disaster.

Someone once said, "Man's flight through life is sustained by the power of his knowledge." Nowhere is that more true than in the case of an alternator failure during single-engine, night-IFR operations. However, before multiengine pilots stop reading this, let me pass on a little story.

In a recent discussion with the pilot of a medium-sized corporate twin, I was extolling the virtues of having two engines, and therefore, two alternators. I pointed out the incredible odds against ever suffering total electrical system failure. He smiled and mentioned that the previous month, both of his alternator clutches failed simultaneously—fortunately it happened while he was still on the ground. Two-engine types beware, because it is not outside the realm of possibility.

During the days when aircraft had few or no electrical systems, more often than not, the pilot was also the mechanic. Airplanes were mechanically simple and the solution to a given problem was fairly obvious. As demand increased for more sophisticated systems, being a pilot began to require increased development of flying skills, leaving less time to devote to mechanical familiarity.

It is not hard to imagine that as electrical systems became more involved, mechanics began sketching them out before actually wiring the airplane. The more complex the systems became, the more careful the planning and the more elaborate the schematic drawings. Because these schematics were intended to be used by engineers and mechanics, no attempt was made at realistic depiction of components. Ultimately systems became so complex, problems could arise that were no longer easily solved or even understood.

Pilots became aware of the need for increased systems training to be able to troubleshoot problems in flight. Because schematics had become the key to unraveling the mysteries of any given system—not just electrical—pilots had to learn to interpret them. Unfortunately, systems continued to become more complex, and schematics rapidly took on nightmarish qualities. I cringe when I think of the electricity course I had to take as an A&P student at the University of Illinois and the hours spent pouring over incomprehensible electrical system schematics.

ELECTRICAL SYSTEM SCHEMATICS

The purpose of a schematic is to provide a means for the mechanic to trace a system visually; it is typically used as an aid in locating the source of problems and essentially is a road map for troubleshooting. Unfortunately, standard schematics go into far greater detail than is usually desired by the pilot; from his or her point of view, only information that can be used to solve inflight problems is useful.

Thanks to the General Aviation Manufacturer's Association (GAMA), simplified pictorial schematics were initiated a few years ago for general aviation aircraft. The effect was a quantum leap forward in pilot understanding and ability to troubleshoot systems. Gone were the obscure symbols for motors, alternators, starters, switches, and other components. Everything was replaced by simplified, miniature drawings of actual components: alternators looked like little alternators, master switches like little master switches (FIG. 16-1). Fuses, circuit breakers, switches, and other controls and equipment are easily distinguished and diagrammed in a simplified, functional order that enhances system understanding. The modern pilot-oriented schematic permits good system understanding by even the most mechanically unsophisticated pilot.

When I was a King Air simulator instructor for Flight Safety International, Inc., Dan Orlando (whose expansive knowledge earned him the nickname "Mr. King Air") explained to me one of the most valuable little training aids I ever have received in aviation. As Chief Pilot for King Air Learning Center, Orlando was not only concerned about the crews during their training program but also about maintaining their proficiency between training sessions. He encouraged me to introduce a game called "What If" to all my crews.

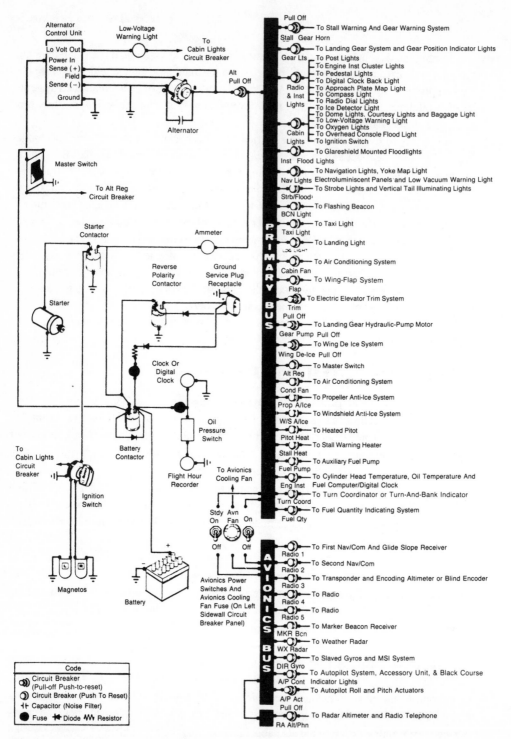

Fig. 16-1. *Electrical system schematic.*

138

Designed to keep a pilot sharp on systems and emergency procedures, "What If" can be played anywhere. If there are two or more players, one person asks a question and the others see if they can answer it. For instance, "what if the gear won't extend, even with the manual gear-extension procedure?" Or "what if the alternator field circuit breaker pops?" If you are playing alone, then sit in your airplane with the manual close at hand and ask yourself "what if" questions, referring as necessary to the procedures and system section of the manual.

To be able to play "What If" successfully, a pilot must be familiar with the systems and procedures of the airplane. "What if-ing" the electrical system requires not only a basic understanding of the system but what specific equipment it operates. Radios, lights, and pitot heat are obvious, but there are some not-so-obvious ones. Some equipment may or may not use electricity, for example hydraulic landing gear, fuel valves, gyros, and stall-warning devices. On the other hand, some instruments that you might expect to be powered by the electrical system might not be, such as cylinder head temperature, oil pressure, and oil temperature gauges.

SYSTEM FAILURE

Electrical system failure is a very commonly misused phrase that almost always actually refers to alternator or voltage regulator failure. In such cases, what happens is that the alternator no longer is capable of providing electricity to the electrical system and the battery takes charge. The electrical system still functions, but at the cost of an ever-discharging battery. There is no way to recharge the battery, and if the situation continues long enough, a real electrical system failure could occur. However, actual instances of total electrical system failures are almost nonexistent.

By way of introduction to the subject of alternator failure, I asked my class to tell me what they felt the average pilot thinks about alternator failure. One student quickly responded, "They know it's not a problem on day-VFR flights." The observation was good, unfortunately the premise was not. Contrary to popular belief, all inflight alternator failures are bad. I agree there are degrees of bad, and probably the worst condition would be night IFR flight in icing conditions, but day VFR also can present some sticky situations. Besides avionics equipment, there are landing gear and flaps to consider; you can usually manually extend the gear, but what happens if you have to go around and there is no power to retract? Night flight, however, does present a major problem. Take, for instance, the simple challenge of being able to see whatever instruments still are working.

I had an instrument student who was of the opinion that if the lights went out in flight, he would simply hold his flashlight in his mouth. While on a dual flight one night, I turned off the battery master and informed him that he had "lost" the electrical system. Smugly, he reached over and pulled a flashlight out of his flight bag; one of those nice, big, bright, ribbed chrome types. It just barely fit in his mouth, took about one minute to make his jaws ache, three minutes to make his head stoop, almost knocked out a tooth when we hit a little bit of turbulence and the light intensity was blinding as it reflected off the instruments and windshield. Could you imagine doing that for an hour or two?

The simple truth all pilots must be willing to face is that you have to be prepared for the possibility of alternator failure. What makes this failure such a problem is that the electrical system continues to operate normally as if nothing had happened; the battery immediately takes over the power requirements of the system. Due to the failure of the alternator, the battery cannot get recharged, so any electrical demand is a drain on the battery. The greater the demand, the faster the battery will become exhausted. The trick is to identify the problem immediately, while the battery is still fully charged, so you can assure sufficient power for the remainder of the flight. That is sometimes easier said than done. If the electrical system itself doesn't show any change initially, how does a pilot know when an alternator failure occurs?

TROUBLESHOOTING

The prime source of information regarding alternator failure is the ammeter. The telltale indication depends on which of two types you have in the airplane. The zero-center type of ammeter shows charge/discharge rate; when it shows a discharge you can assume the alternator is inoperative. The other type of ammeter is the zero-left or loadmeter type, which indicates the amount of load the alternator is supplying. A zero indication of the loadmeter indicates the alternator is inoperative. (See Chapter 13: Aircraft Alternators: the Current Solution).

The ammeter is seldom in the pilot's normal instrument scan and often is installed on the far side of the instrument panel or somewhere below the pilot's normal field of vision and obscured by the yoke. Many aircraft have a red warning light to indicate alternator failure which is certainly helpful, but bright daylight conditions and preoccupation with other problems have both been known to distract a pilot sufficiently enough for the light to be overlooked. The safest course of action is to learn to include the ammeter in your instrument scan.

It is important to point out that an inoperative alternator does not necessarily indicate alternator failure. In the event you have an inoperative alternator indication, before taking any drastic measures (such as load shedding or early termination of the flight), do a little snooping around. It is possible that the master switch has accidentally been shut off, or the alternator (or alternator field) circuit breaker has tripped—a reset attempt might be successful. In aircraft that have the on/off type circuit breaker installed for the alternator, it could have accidentally been shut off.

A good rule of thumb in all troubleshooting is to initially ask yourself what is the most obvious possibility for the failure? Often the answer is "mis-control" of the system, such as accidental deactivation. While particularly distasteful to accept, it is nonetheless very commonplace for pilots to turn off master switches instead of turning off lights, to switch to empty fuel tanks instead of full ones, and to raise landing gear after touchdown instead of flaps.

Most pilots should agree at this point that inflight alternator failure is a nasty situation at best. While the importance of early recognition can hardly be disputed, it does not assure the safe outcome of the flight. It is early recognition combined with knowledge of alternatives that assure the best possible decisions under the circumstances.

LOAD-SHEDDING DECISIONS

To make appropriate load-shedding decisions, the pilot should study the electrical system schematic while on the ground to get an understanding of how the system is designed and what can go wrong. The more complex the system, the greater the need for digging in and finding out what makes it tick; however, even a relatively simple system merits a thorough review and some research.

Looking again at the sample schematic of FIG. 16-1, it is possible to quickly note several important details. This system has a "split master," which allows the pilot to easily remove the alternator field from the system. This is important, because even if the alternator fails, the alternator field continues to demand precious current from the battery. Therefore, when the alternator fails, the pilot can and should shut off that portion of the master switch. It also shows which items have circuit breakers so the pilot can remove these items from the primary bus. These include the landing gear hydraulic-pump motor, strobe lights, and alternator, as seen in the example. In addition, there is an avionics power switch (avionics master), that permits separation of all avionics from the primary bus, a valuable practice during engine start to reduce the risk of damaging avionics from possible, potentially harmful voltage fluctuations. Another useful piece of information is knowing exactly which items are protected by each circuit breaker. If, for instance, the circuit breaker marked *NAV LT* trips, the problem could be rooted in four possible areas: the navigation lights, the yoke map light, the electroluminescent panels, low-vacuum warning light, the wiring to any of those items, or the circuit breaker itself. Experimentation will produce sufficient information quickly enough for you to make intelligent decisions.

First, turn off all the items under the control of that circuit breaker. Then after a two-minute cooling period, reset the circuit breaker by pushing it in. Wait for a minute or two, and if it pops again, then the problem is either in the wiring or the circuit breaker itself—both beyond the pilots inflight capability to repair. If nothing happens, activate one of the systems and wait another minute or two. If nothing happens, turn it off and activate a different system and wait. Chances are pretty good that eventually you will activate the faulty system, causing the circuit breaker to trip a second time. It is worth noting that a glitch in a system occasionally causes a circuit breaker to trip, so upon resetting, the breaker will not trip a second time, and so much the better. This procedure should be followed whenever there's a trip of any circuit breaker that protects more than one piece of equipment.

What constitutes essential equipment? "Non-essential equipment" as it applies to the electrical system refers to equipment that should be turned off in the event of alternator failure. CFIs, when discussing alternator failure, tend to say, " . . . and of course then you turn off all non-essential equipment." Unfortunately many CFIs never actually get around to telling you how to determine what is non-essential.

Certainly, if you are flying along listening to stereo music and have a portable coffee maker plugged into the cigarette lighter outlet, there is some fairly obvious non-essential equipment being used. How about the situation where you are flying IFR at night and the closest airport with weather good enough for an approach is more than an hour away? The solution to the problem is to manipulate usage of the electrical equipment to assure you will not exceed the capability of the battery for the time you anticipate using it. That means you will have to begin thinking of each piece of equipment in terms of its necessity at any given moment, as well as how much amperage it will require to use it.

Take, for instance, a battery rated at 40 amps per hour. Theoretically, it will support a continuous system load of 40 amps for 1 hour. If you expect your flight to last 1 hour, you have 40 amps at your disposal during that hour. That implies you can use whatever equipment you want, provided you never exceed a total of 40 amps demand at any given time for 1 hour. Remember this is purely theoretical; in reality, things just don't work that neatly. For one thing, it presupposes the battery was fully charged and in perfect condition. It also does not take into consideration that some equipment, such as radios, might not work as the battery becomes significantly discharged, even before the anticipated time expires. Therefore, in this case, it is not practical to think that reducing system demand to 40 amps will guarantee 1 hour of use. It always is a good idea to reduce the load as much as is safely possible and to terminate the flight as soon as you can.

DEVELOPING A LOAD SHEDDING CHART

To make intelligent decisions about what equipment to use and what to turn off, the pilot needs to do a bit of homework. One afternoon when you have the urge to fly but the weather isn't agreeable, take the electrical system schematic to the airplane, find the current draw for each item listed, and develop an electrical-emergency load-shedding list like the one depicted in TABLE 16-1 for a common general aviation single-engine aircraft. This can be done in a number of ways. Perhaps the easiest is to check the circuit breaker rating for each item. While the rating is higher than the actual current draw for that particular item (approximately 1.1 to 1.5 times higher), it gives you a pretty good estimate on the safe side. A more exact method, though more time consuming, is to get the rated amperage from the data plate on each appliance (the same information also can be found on manufacturer's information sheets). Compare the differences in the circuit breaker amperage ratings shown in the electrical-emergency load-shedding list and the list of actual rated amperage for the same airplane (TABLE 16-2). It is worth noting that some older communications radios might require up to 2 amps to receive and 6 amps to transmit.

Table 16-1. Electrical Emergency Load-Shedding List

Amps	Item
20	Taxi and landing lights
20	Landing gear hydraulic-pump motor
10	Pitot heater
10	Wing-flap system
10	Flashing beacon
5	Auxiliary fuel pump
5	Avionics cooling fan and strobe lights
5	Instrument cluster
	Low voltage warning light
	Ignition switch
5	Carburetor air temperature gauge, map, compass, instrument lights and dimmers, dome and courtesy lights, post lights
5	Navigational lights
	Yoke map light
5	Turn coordinator
5	Radio 1
5	Radio 2

Table 16-2. List of Actual Rated Amperage

Amps	Item
17.5	Landing gear motor
9.0	Taxi light
9.0	Landing light
8.5	Wing-flap system
6.0	Flashing beacon
3.0	Strobe light
2.5	Autopilot
2.25	Com (transmitting)
2.0	Transponder
1.2	DME
1.2	Dome and courtesy lights
1.0	Avionics cooling fan
1.0	Navcom (receiving)
1.0	ADF
0.7	Map, compass, and instrument lights
0.5	Glideslope
0.1	Marker beacons

Once you have studied the system and have your electrical-emergency load-shedding list, then you are prepared to make intelligent decisions should you be forced to rely only on battery power. It is easy to see, for instance, that taxi and landing lights should be avoided and the gear should not be extended until you are absolutely sure you will be landing, as the price for extension and then retraction is very high indeed. Take the situation where you are about to fly an instrument approach to an airport with marginal weather. If there is the likelihood of a missed approach, you might seriously consider manually extending the gear to conserve battery power for a possible trip to the alternate airport; the gear then can be electrically retracted in the event of a missed approach.

One important lesson to be learned by all of this digging and probing is that the average battery in good condition should provide sufficient power to run necessary electrical equipment long enough to get to a suitable airport. Not only is that fact important from an operational standpoint, but perhaps even more important from a psychological one.

PART 3
Aircraft Systems

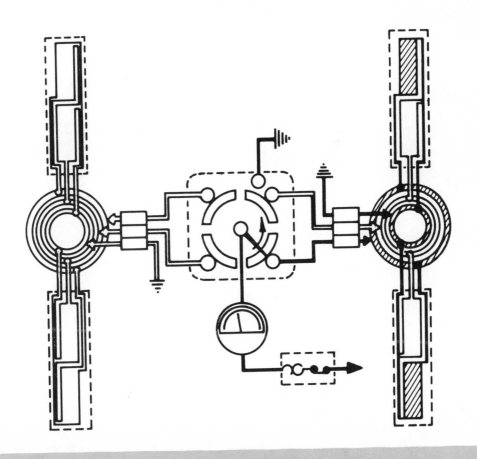

17

How to Prevent Propeller Problems

A propeller can be defined as a rotating airfoil with two or more blades attached to (or integral with) a hub. Powered by an aircraft engine, its purpose is to convert engine horsepower into thrust. Propellers are designed to be compatible with specific engines on specific aircraft and are not interchangeable.

There are two basic configurations for propellers: *tractor* and *pusher*. The tractor propeller, mounted on a forward facing engine, is most common. It's so named because it pulls the airplane along like a tractor would. The pusher prop is mounted on a rear-facing engine and pushes the airplane from behind. One twin engine aircraft, the Cessna 337 Skymaster, uses both a tractor and a pusher prop. The aircraft had twin engine performance with the safety and ease afforded by centerline thrust.

While most light aircraft have two-bladed propellers, three or more blades are very common on high performance singles and multiengine aircraft. There are several advantages to more than two blades. The individual blades can be shorter, allowing increased ground clearance without a decrease in performance. Shorter blades have higher, and less objectionable, sound frequencies and an overall reduction in vibration. The additional blades also produce a greater flywheel effect and generally improve aircraft performance.

Propeller tips can be rounded or squared, depending on noise requirements, blade-vibration characteristics, and other special design considerations. Tests have shown that elliptical tips are slightly more efficient than square but that's not the reason square tips are more common; they're more practical. Square tips leave extra material that can be removed after damage occurs, turning it into a round or elliptical tip for instance, and still maintain the required prop diameter. That's important, because prop diameter (the circle circumscribed by the blade tips) is carefully designed for maximum efficiency. Slower aircraft use a larger diameter while high performance aircraft have smaller diameter props. The longer the propeller blade, the greater the distance the tip must travel during each revolution. High speed aircraft engines can cause the tips of long propeller blades to reach the speed of sound. This results in significant aerodynamic breakdown, vibration, and ultimately prop destruction.

Two terms that frequently confuse pilots are prop *face* and *back*. The flat side of the propeller, that which faces the *pilot*, is called the face; the opposite, cambered side is the back.

Blade angle is the angle between the plane of rotation and the chord line of a given propeller airfoil section (FIG. 17-1). Blade angle decreases along the length of the blade (called *pitch distribution*) with the greatest angle nearest the hub. The reason a prop has pitch distribution is because the tip must travel at a higher relative velocity than the hub. This can easily be observed by tying a key to a length of string and spinning it over your head. The key and string, essentially one continuous object, must turn at the same rpm, but the key covers significantly more distance. This is where the prop most significantly differs with the wing, an airfoil that receives a fairly constant relative wind throughout its span. Pitch distribution attempts to make the prop efficient along its entire length to obtain maximum forward movement. The distance a prop section moves forward in one revolution is called pitch and is measured in inches. The amount of lift produced by any given prop varies with airfoil shape, angle of attack on blade sections, and prop rpm.

Pitch distribution makes talking about a specific blade angle difficult; we need a kind of "street address" to help us identify specific blade angle locations. *Blade station* is a specified distance measured in 1-inch increments from the center of the hub outward. For

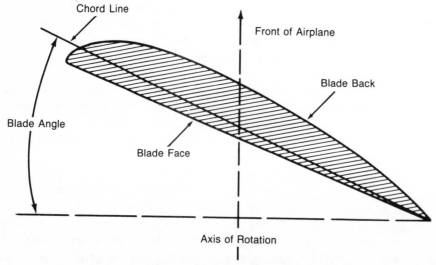

Fig. 17-1. *Cross-section of a propeller blade.*

instance, in one model of the Beech B-55 Baron, the Pilot's Operating Handbook (POH) describes a feathered condition as one where there is an 80.0-degree blade angle at the 30-inch station.

Over the years, several different types of propellers have been developed. The ground-adjustable type found on older aircraft permitted the pilot to set prop pitch on the ground. Normally set to provide maximum efficiency during straight-and-level flight, it could be reset to an angle more conducive to short field takeoffs; the penalty would be slower cruise performance. The problem was, most fields were short in those days, so cruise efficiency suffered. A two-position prop was developed. Then, from within the cockpit, the pilot could switch from low pitch for takeoff to high pitch for cruise. The next logical step was a controllable pitch propeller that allowed the pilot to select any blade angle within the prop's range, but it was a matter of guesswork to decide which was the best. The short-lived automatic pitch prop had the capability of setting its own pitch as a result of aerodynamic forces—the pilot had no control at all!

The fixed-pitch propeller is as common today in light, single- engine trainer aircraft as it was during the heyday of the Wright Brothers. The wooden version of this reliable one-piece prop, long a coveted decoration for the pilot's den, is still providing excellent service in the air. Construction consists of several laminated layers of wood with a doped cotton fabric sheathing glued to the last 12 to 15 inches of the blade (FIG. 17-2). Occasionally a plastic coating is used instead of fabric, but they both serve to reinforce the thin blade tip. A metal tipping, usually made of brass, Monel or stainless steel, runs along the leading edge of the prop, which helps prevent foreign object damage. Small holes are drilled into the metal tip to allow the wood to breathe and moisture to escape. For ease of maintenance and reduced weight, most modern fixed-pitch props are made of aluminum. Fixed-pitch props can be purchased, depending on the operator's needs, for maximum efficiency during climb or cruise. Most aircraft would have the latter installed, but aircraft primarily used for hauling sky divers or operated routinely off of a short, sod strip might do better with a prop designed for climb efficiency.

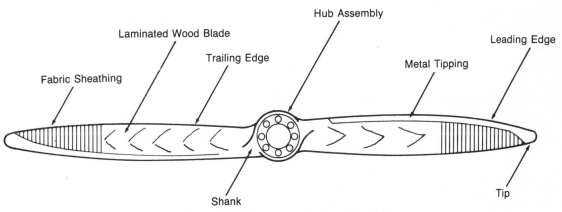

Fig. 17-2. *Parts of a wooden propeller.*

The modern constant-speed prop allows the pilot to select an engine rpm based on current operating conditions. The prop governor adjusts the blade angle to maintain selected rpm. This type of prop is used on most medium and high performance singles and practically all propeller-driven multiengine aircraft. One significant advantage is the ability to reduce prop drag to virtually zero (called *feathering*) should an engine fail in flight. When feathered, the prop blade turns its edge into the wind and the prop comes to a stop. Some aircraft, typically large recips and turboprops, even have the ability to rotate the blade angle to a negative value, which effectively creates thrust in the opposite or reverse direction. Prop reverse significantly reduces landing roll; when combined with differential power, it greatly improves ground handling. For more information about constant-speed props, see Chapter 18: Constant Speed Props: How They Work for You.

PROPELLERS AND THE FARS

The manufacture of propellers is very tightly controlled by Federal Aviation Regulations (FARs). They not only define the limitations of the prop but also its effect on the engine. The single-engine prop must limit engine rpm to the maximum allowable when the engine is at full power and the aircraft is at its best rate of climb speed; this prevents engine damage due to overspeed. This is why you are unable to reach the red line on the tachometer during maximum power ground run-up in a zero wind condition. In addition, the prop must prevent the engine from exceeding rated rpm by no more than 10 percent in a closed throttle dive at the aircraft's V_{NE}. The constant-speed prop must always restrict the engine to rated rpm during normal operations. In the event of governor failure, static rpm must not exceed 103 percent of the rated rpm. This is essentially what determines where the manufacturer sets the low blade (high rpm) angle.

Similarly, after years of practically no standardization, the FARs now detail the design of cockpit controls and instruments. Forward movement of controls produce an increasing effect, so mixtures enrichen, prop rpm gets higher and forward thrust increases. An aft movement of the throttle on an aircraft with thrust reverse will place the prop blade at a negative angle and increase the reverse thrust. It is also required that controls be easily distinguished from one another as to shape and color. Instrumentation must use standardized markings: red radial line indicates maximum operating limitation; green arc, normal operating range; yellow arc, takeoff and precautionary range; and red arc, critical vibration range.

Another critical area covered by regulation is minimum terrain clearance or the space between level ground and the edge of the prop tip. Ground clearance for land aircraft assumes normal inflation of struts and tires. For conventional landing gear aircraft, it must be at least 9 inches in the takeoff attitude; for tricycle gear it is 7 inches in the most nose-low, normal attitude, whether stationary, during taxi, or on takeoff. It also requires positive clearance with struts and tires deflated. For the seaplane, minimum prop-to-water clearance is 18 inches.

PREVENTIVE MAINTENANCE

Probably the best preventive maintenance possible is to keep props clean. Otherwise, it is very difficult to see cracks and other problems while they are still repairable. To wash wooden props, use a soft brush or cloth and apply warm water and a mild soap across the entire blade. When finished, dry with a soft towel. Metal props should be cleaned with a non-oil-based cleaning solvent approved by the manufacturer (Stoddard solvent is a good example). Never use a caustic cleaner or acid. The solvent can be applied with a soft brush or cloth. After wiping the prop dry, wax the blades with a good auto paste wax to protect them from corrosion. For the same reason, it is also a good idea to wipe the blades periodically, even after every flight, with an oily rag.

Propeller repairs typically have to be accomplished by the manufacturer or an authorized repair station. Some may be done by a powerplant mechanic, but there are no repairs authorized for the non-A&P. The owner/operator is primarily limited to basic preventive maintenance with replacing defective safety wire or cotter keys. It is permissible to lubricate parts of a propeller not requiring disassembly beyond non-structural fairings, coverplates, and cowlings. You may also apply coatings such as paint, wax and other preservatives if they are not prohibited by the manufacturer and are not contrary to good maintenance practice, and then only if no disassembly is required. Major alterations and repairs may only be performed by an authorized repair station according to Advisory Circular (AC) 43.13-1: Acceptable Methods, Techniques and Practices. This would include repairs to deep dents, scars, cuts, and nicks. Tolerances are tight, with little room to work. For instance, take a prop with a diameter of less than 10 feet 6 inches. A nick located between the hub and the 24-inch station could not exceed $3/64$ of the blade width. If the blade width was 6 inches, a nick deeper than approximately $1/4$ inch would be cause for a repair facility to scrap the blade! But if that seems unreasonable, consider the forces acting on the propeller.

Vibration is a natural by-product of the propeller as it produces thrust, but excess vibration work-hardens the metal and leads to failure. Mechanical vibration, the result of piston engine power pulses, is a prime culprit in metal fatigue and prop failure. Some rpms are particularly harmful to the prop, so the manufacturer puts a red arc on the tachometer. Operation is not permissible in the red arc except to pass through it. For this reason, it is important to have the tachometer checked for accuracy every 100 hours.

Thrust, aerodynamic twisting, centrifugal force, and torque are operational forces that act upon the propeller. With so much working against the propeller, there are amazingly few failures, and those that happen can almost always be traceable to fatigue cracks resulting from nicks and other scars left unattended. Therefore, because of the potential for problems more so than the likelihood, manufacturer-prescribed routine maintenance should be carefully adhered to. This is especially true for manufacturer-recommended overhaul of the constant-speed prop. Typically based on hours in service, it includes complete disassembly, inspection, reconditioning and replacement of parts as necessary, and re-assembly.

Despite years of faithful service and their aesthetic value, the wooden prop does have a wide range of potential problems. In particular, delamination (separation of laminations) is cause for hanging it in the den; repair is possible if only the outer lamination has begun to separate. Dents and other scars indicate real trouble in wood because they indicate cracks, and cracks lead to failure. If caught in time, small ones may be repaired with an inlay. Other minor defects might be curable with filler; small cracks, parallel with the grain, may be stopped with resin glue. Tip fabric should always lay neatly and metal tipping should be smooth and uninterrupted—if not, consult a repair station as these problems tend to worsen rapidly. Cracks in the solder joints near the metal tipping can indicate wood deterioration beneath and always merit a thorough inspection by a mechanic.

Some defects are beyond repair. They include: a crack or cut across the grain, elongated bolt holes, warped blades, nicks or chips with significant wood missing, oversized crankshaft bore, a split blade, and cracks between the bolt-attach holes.

Aluminum alloy propellers have many advantages over their wooden equivalents. They are thinner with equal or greater strength and often weigh less in the one-piece construction type. Leading and trailing edge defects can be dressed out by a powerplant mechanic, provided the finished size is less than $\frac{1}{8}$ inch deep and 1.5 inches long and the repair has smooth and gradual curves. A slightly bent blade can be repaired, but there are precious few degrees of freedom.

CORROSION

The snake-in-the-grass for aluminum is corrosion; the owner should do all he can to ensure against its insidious effects. Corrosion forms tiny cavities that can extend deeply inward, tunneling under the surface of the propeller and reappearing elsewhere, like chemical wormholes eating your airplane. Props with deice boots and leading edge abrasion boots should be carefully checked for corrosion during 100-hour inspections, because the damage might be hidden. Corrosion is especially a problem for aircraft routinely operating in high moisture or salt water areas, and these aircraft should receive more care. Never attempt to remove corrosion with steel wool, emery cloth, steel wire brushes (except stainless steel), or severe abrasive materials. Particles of steel wool or emery cloth will become embedded in the aluminum and lead to an even greater corrosion problem.

PREFLIGHT

Preflight of the propeller should always begin with checking to make sure the mags and master switch are *off*. Even so, always stay out of the prop arc during your preflight; never lean over the prop. After visually inspecting for corrosion, while standing clear, grasp the blade tip and test for looseness by pulling fore and aft. Some movement, called *blade shake*, is normal for constant-speed propellers because of their design. Run your hand over the face and back of the prop and your fingernail over the leading and trailing edges searching for irregularities of any kind. Nicks in the blade radiate lines of force outward, causing irreversible damage. If small and caught early enough, they can be dressed by a mechanic. If a small nick occurs while the airplane is in a location where no mechanic is available, as an emergency measure only you may very lightly dress it with a fine, half-round file, just enough to smooth the jagged edge. Do not use emery cloth, because particles will become embedded in the soft aluminum material and lead to corrosion. Then fly the airplane to the nearest A&P; if in doubt about its severity, don't fly it. Blade failures typically occur within inches of the prop tip, but there have been cases of failure as far inboard as the hub, so don't be fooled.

Contrary to popular belief, most prop spinners are not optional items. They are typically used to assure smooth cooling airflow into the cowling. Spinners should be checked for cracks, security of attachment, and evidence of oil. Oil indicates crankshaft seal failure. While standing near one end of the prop, sight down one tip across the blades to the other tip and check for alignment. If there appears to be a bend, you should perform this simple blade track check that assures one blade tip follows the other in the same plane.

After double-checking mags and master to be sure they are off, point one prop tip directly down. Block up a smooth board directly under the prop tip and pull gently on the tip. Mark the spot on the board with a pencil, push the tip gently backward, and mark that spot with the pencil. Without moving the board or the airplane, rotate the other prop into the down position and repeat the procedure. The two sets of marks should not vary by more than $\frac{1}{16}$ of an inch. If they do, have it checked by a mechanic.

Safe operation of propellers is mostly a matter of common sense. Without a doubt, the prop is potentially the most lethal part of an aircraft, even when the engine is not running! Connected directly to a reciprocating engine, moving the prop means turning the engine over. That includes turning the mags and pulling air through the carburetor. Any number of system-related failures ranging from a broken magneto P-lead to a fuel leak could cause the engine to start unexpectedly. Turning a prop by hand, even backwards, should always be considered unsafe and avoided if possible. The old routine of putting a prop in the vertical position to indicate a refueling order is an accident looking for a place to happen—use a windshield sign instead. Pulling an airplane around by its propeller is also not a good idea; it's hard on the blade when done at the tip and hard on the crankshaft and bearings when done at the hub, and of course there is always the potential for accidental engine start. Use a towbar. It's much less expensive and time consuming than an airplane out of control on the ramp. Another worthwhile rule of thumb is to never get near a prop with a non-pilot in the airplane. There have been too many instances where a passenger has accidentally turned on the mags, pushed a mixture lever forward, or even hit a starter.

There are few nevers in aviation, but an engine should never be started when there is an unattended small child in the area. It's really a good idea to avoid starting whenever there are people in general nearby. Always shout "clear," "prop" or whatever your fa-vorite term is immediately prior to actual engine start, regardless of where the checklist puts it. Then look around the aircraft before engaging the starter—to the uninitiated, "clear" might make them look up to the sky and smile unknowingly. Even with all these precautions, some people can be mesmerized by a rotating propeller. Some years ago, a pilot's wife walked back to the airplane for one last goodbye; she walked right through the running prop and lost an arm. To help reduce the potential for blindly walking into a prop, it is a good idea to paint the back highly reflective colors such as a red tip, yellow stripe and then red stripe. The face of the blade should be painted flat black to prevent the pilot from being subjected to strobe effect. This problem occurs when flying with the sun at your back and at night with the aircraft lights on.

If, while in flight, you have reason to believe there is an impending blade failure, reduce the power to idle. The slower rpm will help you observe the prop to assess the situation. If there is an actual separation, immediately shut down the engine, as the resultant vibration could easily tear the engine from its mounts. While the propeller is a highly reliable piece of equipment, failures do occur. It is interesting to note that the first fatal aircraft accident killed Lt. Thomas Selfridge while he was observing Orville Wright on the Wright's "Military Flyer." Occurring on September 17th, 1908, it was the result of a prop failure.

18

Constant-Speed Props: How They Work for You

A constant-speed propeller system is one in which propeller blade angle is varied by a governor, so a constant propeller rpm can be maintained, despite engine throttle changes or variations in aircraft speed. The constant- speed propeller can be operated at a high level of efficiency. Unlike the fixed-pitch propeller, which is efficient only at cruise—or in some models, only during climb—the constant- speed prop permits both efficient climb and cruise (FIG. 18-1). For any given phase of flight, there is one rpm setting that is most efficient; it gives you the greatest gain for the least amount of energy. The constant-speed prop allows you to operate, within limitations, at that rpm.

For a propeller to turn at a constant rpm as power conditions change, the angle of attack of the blades must be adjustable. Because it would be impractical for the pilot to make such frequent adjustments, it is done automatically by the propeller's constant-speed governor.

Fixed force and *variable force* are the two opposing forces that control blade angle. During operation, a fixed force tends to either increase or decrease blade angle, depending on the particular design. This fixed force can be caused by centrifugal force acting on counterweights, by a spring, or simply by centrifugal twisting moment. Variable force, which causes the blade angle to change by counteracting the fixed force, is actuated by a governor that controls oil pressure on a piston in the propeller dome. The piston is connected to the blades through a mechanical linkage; this linkage converts the linear, hydraulic motion into a rotary motion necessary for changing the blade angle.

In certain McCauley propellers, for instance, oil pressure on a piston is used to increase blade angle. When the governor diverts oil away from the piston, centrifugal twisting

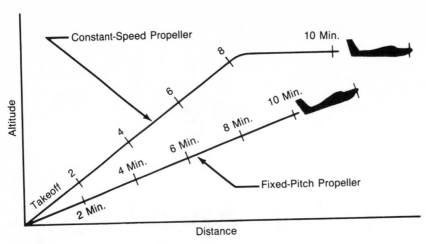

Fig. 18-1. *Climb comparison.*

moment on the blades and a booster spring in the prop hub cause the blade pitch to decrease. Other systems work just the opposite, but the principle is the same.

The governor, connected by a drive shaft to the engine drivetrain, senses engine rpm and compares it to the rpm the pilot has selected with the propeller control lever. It redirects oil pressure to or from the propeller dome, which changes the blade angle to maintain that rpm. An oil pump drive gear located on the drive shaft meshes with an oil pump idler gear and squeezes oil outward, boosting the engine oil pressure to that required for the propeller system. A pressure relief valve bypasses excess pressure to the inlet side of the pump. Oil goes through the hollow drive shaft to the pilot valve on the governor, which moves up or down inside the drive shaft to direct the oil through different ports (FIG. 18-2). One port diverts oil pressure to the propeller dome; another allows it to flow back, relieving the pressure.

Pilot valve position is controlled by a set of fly weights mounted at the end of the drive shaft. The fly weights tilt outward as rpm increases, inward as it decreases. In the outward position, the fly weights raise the pilot valve; in the inward position, they lower it. Pressing down on the pilot valve is a speeder spring, which is connected by control cable, pulley, and speeder rack to the cockpit propeller lever. If the pilot wants a higher engine rpm, the propeller control lever is pushed forward. This compresses the speeder spring, which in turn pushes down on the fly weights, tilting them inward. The propeller, which has been turning at a lower rpm than is now desired, is in an "underspeed" condition. As the fly weights are tilted inward, they move the pilot valve down, permitting oil to flow under pressure to the propeller dome. The oil pushes on the piston, which in turn decreases the blade angle. The lower blade angle allows the propeller to turn faster under the given conditions, so rpm increases. When it does, centrifugal force on the fly weights

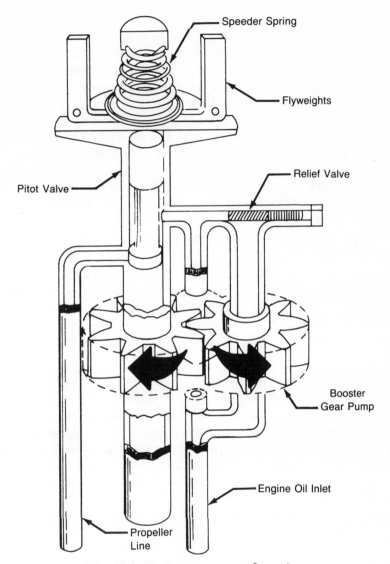

Fig. 18-2. *Basic governor configuration.*

slowly overcomes speeder spring force, and the pilot valve returns to the neutral position. This stops the oil flow and maintains a constant blade angle.

If the pilot wants a lower rpm, the propeller control is moved aft, which relaxes speeder spring tension, tilting the fly weights outward. This condition, called "overspeed," occurs whenever existing rpm is higher than that selected with the propeller lever. The result is the fly weights raise the pilot valve, permitting the oil to flow out of the propeller dome, causing the blade angle to increase, thereby causing rpm to decrease. Decreased rpm results in decreased centrifugal force on the fly weights; slowly, the fly weights again succumb to speeder spring force. The pilot valve returns to the neutral position, the oil flow stops, and blade angle remains constant until the next disturbance. If this sounds involved and complicated, then consider that it occurs almost instantaneously. Overspeed and underspeed conditions are corrected so quickly, the pilot typically is unaware they have occurred and they don't even show up on the tachometer!

The governor can maintain a given rpm only if enough horsepower exists. For instance, every airplane can outclimb its maximum sea-level rpm if allowed to go high enough; normally aspirated aircraft will do so quickly. But there is another source of horsepower available to the propeller besides the engine: free- stream energy. The governor doesn't care if the engine turns the propeller or vice versa, as long as it does it at the correct rpm! If the prop is set for 2200 rpm, the governor will try to maintain that rpm, regardless of throttle setting, by adjusting the blade angle as necessary.

Taken to the extreme, the governor will eventually run out of blade travel, where the prop functionally becomes "fixed pitch," with rpm directly affected by throttle setting (manifold pressure). But, a constant-speed prop has sufficient torque at cruise to maintain the rpm on a dead engine! That is why it is practically impossible to identify visually which of two or more reciprocating engines have failed in cruise flight. The windmilling propeller will turn the engine and its accessories— the generator, fuel pump, vacuum pump, air conditioner, everything. No wonder there is such an incredible increase in drag!

PROP FEATHERING

Engine failure in a single-engine aircraft always means landing, but with two or more engines, that is not necessarily the case. To make the airplane flyable on one engine, it is necessary to reduce drag to a minimum. The solution is to turn the edge of the windmilling propeller blade into the wind (FIG. 18-3). The procedure, called *feathering*, sets the blade angle to approximately 90 degrees, which stops the propeller in flight. The pilot may feather a prop by pulling the appropriate lever all the way back to the low rpm setting, through the safety detent, and into the full-aft, feathered position. Some aircraft, particularly large, older ones such as the DC-3, have a pushbutton control rather than prop-lever detent, but the result is the same. Feathering is totally independent of, and overrides, the constant-speed operation. In some aircraft, the propeller doesn't even have to be turning to be feathered.

Preflight procedures from the pilot's operating handbook (POH) should be followed closely. It is important to exercise the prop during preflight, especially on cold days when the oil tends to congeal. To do this, run the engine up to the recommended manifold pressure with the prop control in its full forward position. Then, quickly pull the prop lever all the way back to the feather detent, get about a 500-rpm drop, and then push it forward again. The entire action should be fairly smooth: pull - drop - push. If the oil is cold, you may not get an rpm drop immediately, so push the prop lever back and forth fairly rapidly to get warmer engine oil moving through the lines. Once you begin to see a response in rpm, continue with the normal prop feather check. Most manufacturers

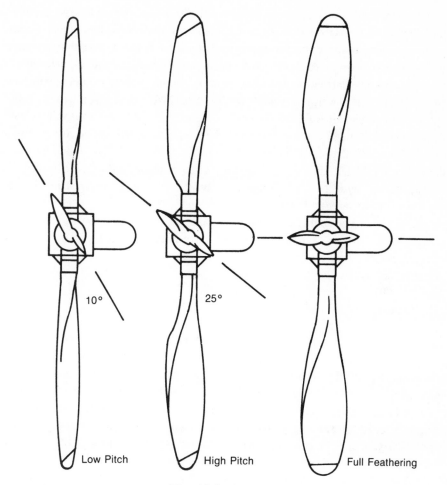

Fig. 18-3. *Blade angles.*

recommend you periodically do a complete feather procedure just to make sure everything will work should you need it. Consult your POH.

The McCauley feathering system, common on many Cessna and Beech twins, uses counterweights and an internal spring that overrides oil pressure to feather the prop. That provides an inherent safety feature: if you lose oil pressure, the propeller automatically feathers. Not a bad idea considering an engine without oil will freeze up anyway! A spring-loaded latch automatically engages during normal engine shutdown to prevent the prop from feathering.

When you feather a prop for practice, there are two ways to unfeather it in flight. Some aircraft, especially those used primarily for training, have unfeathering accumulators. This cylinder contains a diaphragm or piston that separates an air (or nitrogen) precharge from system oil. The pilot moves the prop lever out of feather to the full-forward position.

When the unfeathering accumulator is activated, the high-pressure oil charge is moved to the propeller dome cylinder and forces the blades to a lower blade angle. Once the blade angle has been changed, the force of the relative wind begins to turn the prop (with a dramatic increase in drag). At that point, fuel can be introduced and the magnetos turned on; the engine should start normally. Remember the engine oil and cylinders will have cooled; give them time to warm up at a low power setting before returning to cruise power.

An aircraft without an accumulator requires the pilot put the prop lever full forward and get the engine turning by using the starter. As the engine moves the oil, the governor builds up oil pressure that overcomes the force of the spring and moves the blade to a smaller angle. While this sounds easy, there are a few potential problems. For one, it's more work for the starter; but more significantly, the aircraft is totally reliant on the electrical system and starter. If any problem develops with either of those systems once the prop is feathered, you could have a real emergency on your hands. Hartzell feathering systems are used on Piper, Aero Commanders, some Beech, and older Cessnas. While there are differences, they are fundamentally similar to McCauley in both the feathering and unfeathering modes.

PROP CONTROL

Fixed-pitch props are directly controlled with the throttle; the reference instrument is the tachometer, which indicates engine revolutions per minute. A constant-speed propeller actually is controlled by a propeller lever in conjunction with the tachometer. The throttle controls engine manifold pressure and is referenced by the manifold pressure (m.p.) gauge, which indicates inches of mercury. With the constant-speed system, setting power correctly is essential to efficient operation and even engine life. A pilot always should consult the POH for specific combinations of rpm and m.p. for various climb and cruise configurations. Excessive m.p. for a given rpm could cause engine damage. For instance, an m.p. too high for the rpm causes abnormal cylinder pressure leading to high cylinder-head temperatures, detonation, and excessive stress on engine parts. A good rule of thumb is when increasing power, increase rpm first, then m.p.; to decrease power, pull the throttle back to approximately 1 inch below the desired m.p., then pull back the prop lever to the new rpm setting. The reason for the 1- inch difference is because reducing rpm causes the m.p. to rise slightly.

TROUBLESHOOTING

There are significant differences between the various constant-speed propeller systems. Always consult your POH for proper troubleshooting procedures. In general, McCauley non-feathering propellers go to high rpm if there is a loss of oil pressure; the feathering types go to feather in the same situation. If there is damage to the prop-lever linkage, the governor is spring-loaded to high rpm. The reasoning is, if you don't have feather, then the propeller should go into the condition most conducive for a go-around—low blade angle (high rpm). Failure to unfeather in flight when you don't have accumulators probably means there is insufficient starter power to turn the engine enough to build up oil pressure in the governor. This is not an uncommon problem, particularly in older aircraft.

Before attempting a single-engine landing, you might want to try placing the aircraft in a shallow dive and momentarily activating (called *bumping*) the starter. The increased airspeed combined with the engine rotation from the starter might turn the propeller enough to get the oil moving and start the unfeathering process. If you have an accumulator and the prop fails to unfeather, it probably is the result of insufficient or no air charge, an oil leak in the accumulator or hoses, or a ruptured diaphragm, permitting the air charge to leak into the oil. In any case, there isn't anything you can do about it in flight. You will have to revert to using the non-accumulator unfeathering procedure for your aircraft.

Hartzell feathering propellers with counterweights or spring- assist use an air charge to feather the prop. Low air charge is indicated during the preflight feather check and in flight by sluggish or slow rpm control, especially when reducing rpm. There also may be some difficulty in maintaining rpm, but before you curse the system, make sure you don't have creeping throttle or prop levers. Other indications of the problem could be minor overspeed problems, particularly with rapid throttle application, poor rpm recovery under the same situation, and poor synchronization in the upper cruise speed range. If you are in flight when the problem occurs, reduce the throttle and airspeed until rpm control is regained. Be careful not to go below best single-engine rate-of-climb speed for your aircraft. Once you have regained control, increase the throttle slowly to get as much power back as possible without returning to an overspeed condition. Total loss of air charge would prevent feathering of the prop.

Non-counterweighted models work essentially like McCauley systems. Oil pressure increases blade angle; centrifugal twisting moment on the blades decreases it. Counterweighted models use oil pressure to decrease blade angle and centrifugal force on the counterweights to increase it. Be particularly careful to watch for grease leaks near the propeller hub. Common causes are loose, missing, or defective zerk fittings, defective or loose blade-clamp seals, and over-lubrication of blades. Have a mechanic check out any leaks.

The Compact Propeller version with a low air charge would experience improper constant-speed operation, overspeed, and a surge tendency; loss of air charge would prevent feather. An excessive air charge would cause an inability to achieve maximum rpm and could cause the propeller to feather on the ground when the engine is shut down normally.

PREFLIGHT AND RUN-UP

For all types of constant-speed propellers, check during preflight for obvious oil and grease leaks near the blade shanks. Inspect the spinner for security; this is typically not an optional item and must be in place for proper cooling airflow to occur. Look for excessive looseness of the blades, but realize that some play (called *blade shake*) is inherent in the design. Whenever working on or near the propeller, be certain the magnetos and master switch are off, the mixture is in cutoff, and avoid getting into the propeller arc. Always follow the POH preflight procedures. Should you discover your prop is suffering from reduced or lost air charge, use very gentle throttle movements to prevent potentially damaging prop-overspeed conditions.

Good procedures dictate that the pilot avoid run-ups on areas of gravel, stones, broken asphalt, or loose sand, because they cause prop blade erosion. It is prudent to avoid any operations on gravel runways, but if you have no choice, use the following takeoff procedure provided runway length permits: let the aircraft begin rolling while at low throttle; then gradually increase to takeoff power. This procedure reduces the possibility of the prop picking up a stone and nicking the blades. In general, it is a good idea to taxi slowly and cautiously to minimize hazards such as foreign objects on runways, snowdrifts, taxiway and runway lights, and tiedown chains. Always avoid pulling or pushing the aircraft by its propeller. Done at the tip, there is a high probability of bending the blade; done at the hub, you can stress the crankshaft. And there is the ever-present threat of a hot mag starting the engine. If it is absolutely necessary, grab the propeller shank as close to the hub as possible, but remember, one damaged propeller will cost many times the price of a towbar. The dangers associated with a static propeller cannot be overemphasized; a turning propeller is at best a life-threatening environment. Take great pains to avoid loading and unloading passengers with a turning prop; in fact, don't start an engine until there are no people in the vicinity of the airplane.

PREVENTIVE MAINTENANCE

The prop, considered a separate component of the airplane, typically has a manufacturer recommended time between overhaul (TBO) different than the engine. Know its TBO and follow the manufacturer's recommendations. During each preflight, the pilot should check for wear, nicks, dents and other damage. If a problem occurs, it should either be dressed by an A&P or referred to a propeller shop if the damage is too great. Every 100-hour or annual inspection (whichever comes first), the prop should be carefully inspected by an A&P who will also remove the spinner, check the hub parts for wear and damage, check the air charge, and lubricate it as required. It also is a good idea to have the tachometer checked for accuracy periodically and every 100 hours if it has prop rpm red arc—you could be operating in a red arc area without knowing it!

BALANCING A PROP

Under the best of circumstances, props go out of balance with age, primarily from blade erosion. An out-of-balance prop results in vibration that leads to premature part failures, especially the alternator, fuel control, engine wiring harness, and avionics. It also causes oil cooler leaks, broken or cracked engine mounts, exhaust manifolds and turbocharger mounts, and sheet metal cracks in the fuselage and cowlings.

There are two approaches to balancing a propeller: *static* and *dynamic*. To static balance a prop, it must be removed from the airplane and put on a stand in a shop. The problem with this method is it doesn't take into account the effect of the engine, its accessories, bulkhead, and prop spinners. Dynamic balancing is accomplished while the propeller is on the airplane, taking all aerodynamic forces into account. Also, it can be done at different rpm settings. Many shops now use the Chadwick-Helmuth Vibrex Dynamic Balancer. With this system, nothing is removed from the propeller or the engine; it uses a small accelerometer to measure forces produced by the out-of-balance condition at different rpms. The instrument then tells the mechanic how much weight must be added to the hub to balance the prop; the process is analogous to dynamic balancing of tires on a car.

It is recommended that a prop be balanced every time it or the engine is overhauled; some manufacturers also suggest every 500 hours. It's also a good idea to have it balanced whenever you have a cylinder reworked, after any significant repairs to the prop or engine, and at the onset of any unusual vibration. The benefits are immediate and noticeable. The vibration disappears, performance and engine efficiency improve, and you get the long-term benefit of extended component life.

19

Put the Stops on Brake Problems

Stopping the airplane is what brakes are all about. Friction is how the task is accomplished. Orville and Wilbur had the simplest friction brakes—a tailskid that dragged along the ground. Such an arrangement was more than adequate in those days, because airplanes had very slow landing speeds and the "airports" were actually convenient fields. Of course, every time an airplane landed it tore up the ground somewhat, but with so few airplanes, that really didn't matter. Ground maneuvering was a little tricky; every time you wanted to turn, you had to rev up the engine so the propeller would send a blast of air back over the tail, lifting it up. Once the weight was off the skid, the rudder was used to turn the airplane as necessary.

It wasn't long before flying became popular and pilots started thinking of practical uses for the airplane. First, they thought of carrying one passenger, then two, then a package or two, mail, and as a result, airplanes got bigger and heavier. Approach speeds increased, landing distances grew longer, ground maneuvering became more difficult, and hard-surfaced runways began to appear. The need for pilot-controlled brakes became obvious. It was a logical step to borrow the idea from the automobile brake and adapt it to the airplane. And that's just what was done—the standard drum-and-shoe type brake began to appear on airplanes.

Aircraft ground-handling and control improved, but airplanes continued to get heavier and faster. The drawback of the automotive brake, known as *brake fade*, became evident. The drum-and-shoe setup used friction to stop the airplane with an asbestos-lined metal shoe that wedged against a rotating cast iron drum attached to the inside of the wheel. In principle, it worked well; however, as engine power, gross weights, and speeds increased, the demand on the brake increased. In fact, the modern brake has several jobs: It provides sufficient friction to stop the airplane on landing; it holds the airplane during engine run-up; it absorbs the kinetic energy of the wheel and converts it into heat; and it sheds the resultant heat as efficiently and quickly as possible. The old shoe and drum just couldn't handle it anymore; the heat became high enough to cause the rotating cast-iron drum to expand away from the shoe, which reduced brake effectiveness. A better method had to be devised, and after considerable effort, the disc brake was designed.

Probably one of the simplest systems in most light aircraft, the brakes are usually activated by a dedicated hydraulic system. The major components are: toe brakes, master cylinder, hydraulic tubing, disc, and brake housing (*caliper*), which includes the piston and linings, or pads (FIGS. 19-1 and 19-2). The toe brakes are integral to the rudder pedals, though a few aircraft use either a hand brake or independent heel brakes. When using

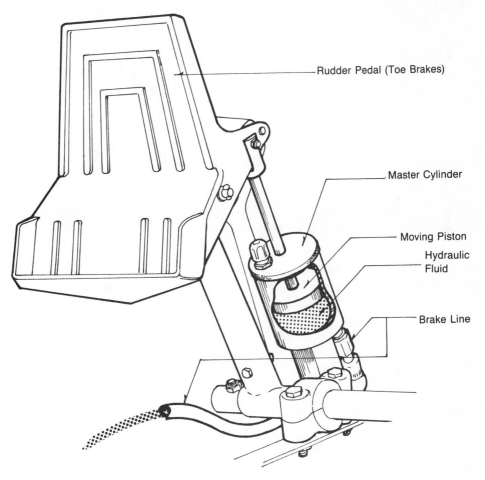

Fig. 19-1. *Standard foot brake.*

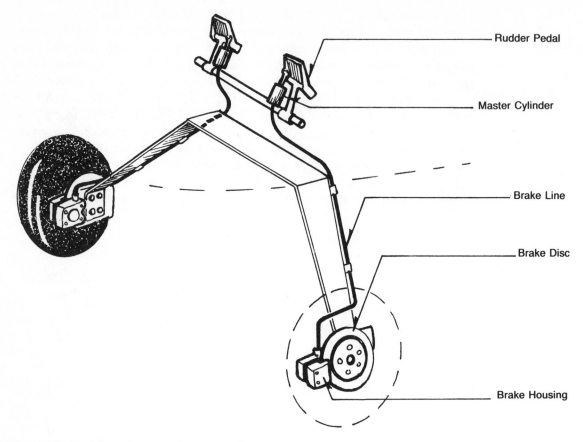

Rudder Pedal

Master Cylinder

Brake Line

Brake Disc

Brake Housing

Fig. 19-2. *Major components of aircraft brakes.*

toe brakes, it is important not to confuse rudder pedal deflection with pushing on the top half of the pedal for brake application. The two functions are independent of each other. The left toe brake operates the left main gear brake while the right toe break operates the right main gear brake; therefore, differential braking is easily accomplished. There is one master cylinder for each toe brake and its purpose is to translate foot pressure into hydraulic fluid pressure through the brake line to operate the wheel brake.

SINGLE-DISC BRAKE SYSTEM

Virtually all modern, light aircraft use a single-disc brake system that is essentially a large, friction-producing caliper—controlled by brake pedals—that clamps on a disc attached to the rotating wheel. Practically speaking, when you talk about light aircraft disc brakes you are talking about Cleveland brakes. According to John Bakos, manager of aftermarket sales for Parker Hannifin's Aircraft Wheel and Brake Division (Cleveland

Wheels and Brakes), 70 percent of all American-built, single-engine and light-twin aircraft come with Cleveland disc brakes as original equipment. The most notable exception is the 100 series Cessnas built after 1974 when Cessna changed to McCauley brakes. The only apparent difference between McCauley and Cleveland brakes is the placement of the bolt holes! Technically, Cleveland brakes are still standard equipment for the smaller Cessnas, and while they don't come fitted with them as original equipment, Cleveland brakes are readily obtainable as replacements.

One reason for the popularity of the single-disc brake system in light aircraft is its simplicity (FIG. 19-3). It consists of a brake unit housing (typically made of aluminum or magnesium alloy) attached to the landing-gear strut and a steel disc that is rigidly fixed to and rotates with the aircraft wheel. The purpose of the disc is to provide a gripping surface for the two brake linings that are mounted in the brake unit housing in a manner that will evenly apply pressure to both sides of the disc. The lining on the wheel side of the disc is stationary, while the lining on the strut side of the disc is movable.

When your foot presses on a pedal, say the right one, the pressure is transmitted by the hydraulic fluid to the piston in the right brake housing. The piston responds to the increased pressure by pushing against the movable lining, which causes friction by pushing on the brake disc, which in turn pushes against the stationary lining, causing more friction. Not only does this provide twice the stopping power, but because of equal friction on both sides of the disc, it also causes even disc and lining wear. The whole process is nothing more than a sophisticated version of the bicycle hand brake.

Disc brakes are usually made of forged steel. For most airplanes that are flown regularly and aren't excessively exposed to a corrosive atmosphere, they are trouble-free. To paraphrase a famous saying, "steel will be steel," and there are some potentially significant problems. If your pride-and-joy flies less than 200 hours per year, and especially if it is exposed to unusual amounts of moisture, salt, or industrial chemicals, you could have a cancer slowly eating away at its discs. If the airplane sits idle for more than 8500 hours per year, not enough time is spent rubbing corrosion and rust off those discs. Not to worry—the folks at Cleveland Wheels and Brake have the answer: chromed discs.

CHROMED DISCS

When most of us think of chrome, we think of the flashy trim on our cars, but chrome discs aren't flashy. Far from being a show item, chrome discs are rather dull in appearance, however they are designed to prevent corrosion and rusting. For a number of years, there were a few mechanics who would take your old, rusted, pitted brake discs, completely resurface them, and finish the job with a chrome treatment. Pilots who used them began swearing by chromed brakes, however there were many skeptics. Most of them imagined the chrome was too slick and wouldn't produce the same braking friction as the traditional discs, but they were wrong. Tests have proven that the coefficient of friction for chromed discs is comparable to regular discs.

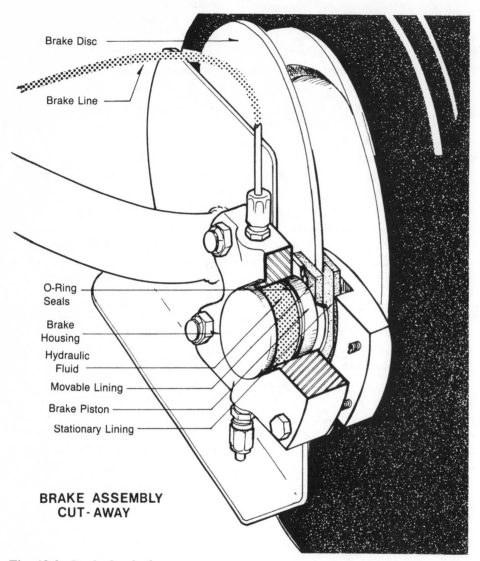

Brake Disc

Brake Line

O-Ring
Seals

Brake
Housing

Hydraulic
Fluid

Movable Lining

Brake Piston

Stationary Lining

**BRAKE ASSEMBLY
CUT-AWAY**

Fig. 19-3. *Single-disc brake system.*

There were some real drawbacks to the refinished disc brakes, though, all of which had to do with the quality control of the resurfacing and chroming process. Today, Cleveland Wheels and Brakes produces the only FAA-TSO'd chrome disc brakes. They start from scratch, there is no refinishing; old brakes are old brakes and are discarded. It's true that a set of chrome disc brakes is more expensive. However, the higher cost is more than offset by the significantly longer life, reduced maintenance costs, and shorter downtime. Most single-engine aircraft now offer the option of chrome discs as original equipment and virtually all singles can have them installed as service replacements.

THE ROLE OF BRAKE FLUID

The prime mover of the brake system is its hydraulic fluid. Without it, there's no stopping you. While it is fairly obvious that brake fluid is responsible for transmitting pressure and energy, it does serve other functions, too. It lubricates the moving portions of the system it comes in contact with and it aids in cooling the working parts.

While there are several types of hydraulic fluid, most light aircraft use a mineral-based fluid that consists of a high-quality petroleum oil. A common type is MIL-H-5606, which is less corrosive than some of the others and can be identified by its reddish color. However, due to its petroleum base, it is flammable and caution should be exercised in using and storing it. It is very important that you check your owner's manual to find out exactly which type your aircraft uses, because mixing different fluids may render the system useless, and some hydraulic fluids can actually eat the rubber seals of incompatible systems! Therefore, it is a good idea to mark the system's filler cap with the type of fluid to be used so no one accidentally adds a different kind. When storing fluid, be particularly careful to protect it from possible contamination by dirt. Particles of dirt can render a system inoperative almost immediately. Try to pick a time to replenish the system when there isn't a dust storm looming up at the edge of the airport—dust and dirt in the fluid are a significant cause of hydraulic system failure!

Even before you start the engine, you can learn whether or not you have sufficient fluid in the system. If you step on the brakes and there is pedal movement, you have fluid; if the pedal lays all the way back and there is no movement, you are out of fluid. Take a look at the floor to see if there is hydraulic fluid on it, an indication of a leaking master cylinder. If there isn't, then step outside and check the ground around the tire; an indication of a leak in the hydraulic line, the brake piston O-ring, or the hydraulic line fitting where it enters the brake housing. Leaks of this nature automatically call for the expertise of a mechanic. Another potential problem easily observed on preflight would be a twisted hydraulic fitting line to the brake housing. If the line has a kink, it will impose a side load on the brake housing and cause uneven lining wear with potentially reduced braking power and shorter life of the linings.

Probably the most dreaded brake problem one can imagine would be to press down on the pedal during the landing roll and get little or no response. Assuming you actually have pedal movement (indicating fluid in the system), the probable cause is dirt. This situation is avoidable if you keep your brakes clean. The culprit is a build-up of dirt on the through-bolts (also called guide pins). These bolts allow the movable lining to slide back and forth in response to the piston. Incidentally, dirt build-up can also cause the brakes to drag (you'll hear a scraping noise when you taxi). This occurs when the brake unit freezes up when the linings are in contact with the disc. To check this during preflight, grasp the brake housing with your hand and try to twist it; if free, it should move slightly; if, try as you may, you cannot get it to budge at all, it is probably "frozen" in place because of packed dirt. While disassembly for cleaning is very simple, it is best to get a little dual instruction from someone who has experience. After you clean the bolts, help prevent the problem from occurring again in the future by lubricating them, but be very cautious what you use for lubricant. Oil, a tempting choice, will only make the problem worse by collecting dirt and holding it in place. Instead, use graphite, Dri-Slide, or silicone spray. All of these lubricate without acting as a sticky surface to attract and hold dirt. Getting into the airplane and stepping on the brakes during preflight might tell you if you have fluid, but it won't give you an indication of the status of the brake linings. The brake pedal will feel fine right up until the second the disc wears off the rivet heads and the linings fall out of the brakes. There is only one way to assure the linings are good; you have to physically get down on your knees and visually inspect both of them. New linings are approximately 0.25 of an inch thick and should be replaced when they are worn to a thickness of 0.10 inch or less. While you are down there looking at the linings, also check for the presence of grease or oil on the linings or disc surfaces. A light coating of either reduces braking effectiveness greatly and could be indicative of a leak.

TROUBLESHOOTING

The Chief Flight Instructor of a major university flight program related a problem to me that surfaced when he started getting a lot of bills for brake relining. Their fleet of aircraft was getting new brakes far too often and the culprit was simple negligence; students were riding the brakes. There is a real potential for that when the rudder pedal serves two purposes. The only cure is to never taxi with your feet flat on the pedal. A better method is to put your feet on the floor and toes on the bottom part of the pedal. It's a simple matter to slide your foot up and press the brake when you need it. If you feel you must keep your feet on the brakes while taxiing to be able to instantly react, you are probably taxiing too fast. Riding the brakes causes unnecessary and rapid wear of the linings. During taxi, you should seldom if ever touch the brakes. Using them to keep your taxi speed slow is an indication of carrying too much power. The automotive equivalent is to pull out of your driveway by flooring the accelerator with one foot and standing on the brakes with the other.

Another common problem is using the brakes to turn, generally a sign of not mentally staying ahead of the airplane (poor planning). Rather than using brakes for turning in the multiengine airplane, there is the advantage of differential power. If you desire to turn left, you just idle power on the left engine and slightly increase power on the right while using left rudder.

OPERATIONAL CONSIDERATIONS

Never raise flaps on the landing roll! That simple rule, ingrained in countless students over the years, has absolutely nothing to do with stopping the airplane. Rather, it stems from the sad fact that many pilots have inadvertently retracted the gear instead of the flaps during landing. Never mind the landing gear squat switch, because they fail. Never mind the gear switch is round and the flap handle is flat—preoccupation tends to cover that up. And never mind they are typically located in totally different places. An alarming number of pilots have overcome all of these obstacles and managed to retract the gear on the runway. Therefore, most instructors tell you never raise flaps on the landing roll!! Unfortunately, that is contrary to maximum performance landings.

Okay, let's be honest. If you are landing your Cessna 152 at Chicago's O'Hare International Airport and you have two miles of runway, who cares? Well, assuming there isn't a DC 10 right behind you, nobody. In fact, it is prudent to use little or no brakes on landing. Given the opportunity, I roll out to the end of the runway to save on brakes. What happens when you are forced to make an actual short field landing? In that situation, everything you have learned about landings and routinely practiced will give you the most inefficient technique. For maximum performance landings, touch down slightly nose-high at the slowest safe speed for your gross weight, with full flaps. This configuration will be effective to approximately 60 percent of your touchdown speed. For example, if your touchdown speed is 100 knots, aerodynamic braking will be more effective than friction down to approximately 60 knots. The reason: the wings are still creating sufficient lift to reduce weight on the tires, which minimizes friction braking. Once you reach 60 percent of your touchdown speed, let the nosewheel contact the ground, retract the flaps, and begin to apply smooth, maximum brake pressure without allowing the tires to skid.

INSTALLING NEW BRAKES

As the saying goes, all good things must come to an end. In the case of brakes, the end applies to linings. If you are fortunate enough to have Cleveland brakes, they are very easily replaced; virtually anyone can be taught how to reline if given the limited instruction, minimal tools, and a half hour of free time. Cleveland brakes are definitely the easiest to work on. You don't have to jack up the aircraft, remove the tire, or disconnect and bleed hydraulic lines. Whether you reline your brakes or an A&P does, when the job is done, they must be conditioned before use. All too often, pilots and even mechanics are unaware of the need for breaking them in. There are actually two conditioning procedures. The one you use depends on the type of linings used on your airplane.

Asbestos-based organic-composition brake linings must have the resins properly cured before the brakes are actually used. Failure to properly condition brakes could result in carburizing the linings (impregnate them with carbon) with a single hard application, preventing a good braking coefficient and significantly shortening the life of the linings. The procedure is very simple. Taxi the airplane at a speed of 25 to 40 mph; then, using a light braking effort, gently bring the aircraft to a full stop. Wait at least 2 minutes for the brakes to cool down. Then repeat. This procedure should be done a total of six times, each time allowing at least 2 minutes for the brakes to cool. Keeping the taxi speed between 25 and 40 mph and using light braking generates sufficient heat to cure the resins but not enough to carburize them. When strictly adhered to, this procedure virtually guarantees properly cured brake linings that should get about 100 hours of taxi time.

The iron-based metallic-composition brake linings require a glazing process after installation. Here a simple procedure is used, with significantly different numbers from the organic method. Taxi speed should be 30 to 35 knots (excess speed can cause overheating and disc warping) with a hard, full-stop braking application. Then immediately go back to the 30-to-35-knot taxi speed and repeat the procedure once more for a total of two times. Unlike the organic break-in procedure, you do not want the brakes to cool down between the two taxi runs. If the procedure is done correctly, the high spots will wear off the linings and the result will be a flat, smooth surface. It is a good idea to check the linings during preflight, and if they begin to appear rough or grooved, repeat the conditioning procedure. One potential problem requiring reconditioning is wearing the glaze off the linings. This is the result of frequent, light brake applications during taxi, or worse, riding the brakes.

PREVENTIVE MAINTENANCE

Hydraulic fluid poses the greatest potential problem. But without it, there are no brakes at all. It is important to maintain the integrity of the lines by making sure they are not kinked or chafing against another part. Routinely check the hydraulic fluid level, and on every preflight, look for puddles of fluid on the ground or streaks of reddish fluid on the airplane. Hydraulic leaks should be referred to a mechanic.

Dirt is the nemesis of brake systems. If allowed to get into hydraulic fluid, even fine grit can cause destruction of seals, erosion of moving parts, and total brake system failure. Allowed to collect around the brake housing and anchor bolts, dirt can cause the brake to "freeze up." Keep the bolts clean and lubricated. One word of caution: do not loosen or attempt to remove the anchor bolts. The procedure for cleaning them is similar to that required to reline the brakes, and while not difficult, it should not be undertaken without some instruction.

Oil or grease on the surfaces of brake discs or linings will cause a significant loss of braking friction; if either is present, remove it with solvent. While you are checking the linings, make sure the surfaces are evenly worn and they are greater than 0.10 inch in thickness. Anything less requires replacement.

With brake system problems, like anything else, you want to give the mechanic as much information as possible. Dragging brakes are indicated during taxi by a squeaking (or scraping) sound from the brake, and if serious enough it might show a tendency for the airplane to pull to one side. When describing problems of insufficient braking, things to note would include whether or not the brake pedal feels "spongy" (indicating possible air in the hydraulic fluid) or if the pedal is flat (indicating no fluid). If you apply even pressure on the brakes and the airplane tends to pull toward one side, you have differential braking that indicates only one of the two brake systems is malfunctioning.

If there is little or no response to the pedal, the mechanic will want to know what the surface conditions were when you experienced the problem. I recall some years ago giving a Grumman Tiger checkout to a CFI who had never flown one. It was a typical Illinois early winter night with a wet ramp and low freezing level. We taxied out, took off, and conducted an uneventful flight. I remember thinking to myself as we entered the pattern, there was a slight crosswind and the pilot had no experience in a caster gear airplane (the Tiger does not have a steerable nosewheel). Because I knew he was a very competent pilot and instructor, I decided to remain relatively docile during the early stage of the landing to allow him plenty of time for self discovery (not to mention recovery). Immediately upon touchdown, the airplane began to veer toward the right, but I remained cool because it was the educationally sound thing to do. I could feel the learning taking place. Well, so much for educational theory, because right about the time he yelled ''you got it,'' we hit the runway light.

I learned something very important that night. When you taxi through water on the ground, and then fly in temperatures above the freezing level, the wet brakes can freeze up solid. The cost of that little educational experience was one runway light, a tire so squared off it would stand upright in a hurricane, and one chief flight instructor with egg all over his face. The end result was rather drastic, but surface conditions can also contribute to brake problems, so they are worth noting.

Low fluid indicates a leak, but unfortunately it doesn't reveal where the leak is located. Puddles of hydraulic fluid discovered during a preflight should be carefully noted with respect to where they are relative to the aircraft. This can lead the mechanic directly to the source of the leak, something that may not be easily found otherwise.

Finally, you should keep track of the condition of the discs. Any of the following irregularities of the disc faces should be reported to the mechanic: warpage, irregular wear, scouring, grooves, pitting, rusting, or corrosion. It can be said that the brakes—perhaps more than any other system—are directly related to the treatment they receive by the pilot. If used correctly, preflighted routinely, and cared for properly, they will give years of useful, surprise-free life.

20

How To Troubleshoot
Tire Problems

"Kick the tires and light the fires" is a half-joking expression that has carried over from the early days of flying. It does point out, however, that tires are an important preflight item, even to the devil-may-care aviator of old. Though dormant from takeoff to landing, aircraft tires can give ground operation a whole new meaning if they don't do their job correctly. They are designed to give a comfortable ride to both the airframe and passengers, to provide easy ground maneuvering, and to maximize braking.

Pilots usually don't think of tires as important control surfaces, but imagine a blowout on takeoff at 75 knots. The immediate and overwhelming loss of control could be devastating, yet tires are probably the most neglected part of the airframe preflight.

Tires are designed and built to be remarkably strong and flexible, so catastrophic failure is not a common problem, though occasionally it does occur. Excessive wear, on the other hand, is the biggest problem with tires.

Ask the average pilot what is the most significant factor reducing tire life and you'll probably get the answer, "Hard landings." There is no question that hard landings make an impression on pilots, passengers, landing gear, and airframes, but hard landings are less of a problem for tires than many pilots think. Aircraft tires are much more flexible than their automobile counterparts and can distort to 32 percent from their original shape. This allows for landings that are a lot harder than the average pilot can handle!

The real killer for tires is heat. The more flexible the tire, the greater the internal heat generated as a result of stress and friction. If you have the slightest doubt about that, drive your family car a half mile down your street at 60+ mph and then while the officer is writing the ticket, feel the tires. The heavier the vehicle, the more heat that is generated, and the more heat, the shorter the life of the tire.

There are several ways to minimize the heat problem, but they tend to be unpopular with pilots:

1. Always keep tires properly inflated.

2. Always taxi as slow as practical; the slower you go, the less the stress and flex and the less heat buildup.

3. Keep ground maneuvering to a minimum: when possible, plan to land in such a way that you will roll out close to your destination on the airport.

4. Reduce—and if possible, eliminate—braking. Not only does this save tires, but it also saves brakes. Braking causes friction between the tire and the ground, which leads to a fast buildup of heat. Also, the brake itself creates a significant amount of heat, part of which will transfer to the tire. A touchdown and long rollout without brakes is preferable to a short ground run with brakes, assuming there are no obstructions looming at the end of a short runway.

There is yet another reason to avoid hard braking: tread wear. Friction means something has to give, and tires are much softer than runways. Every time you apply brakes—airplane or car—you scuff off some of the rubber tread. Tires with little or no tread must be replaced (FIG. 20-1). Also, hard braking leads to skidding.

The reason the wheels don't lock up immediately when you apply the brakes is that the momentum of the airplane combined with ground friction on the tires is greater than the force the brake applies to the wheel. If you touch down on a runway with areas of low friction—such as wet spots or patches of ice—and attempt hard braking, one wheel will eventually cross an area of lower friction than the other. When that happens, the wheel encountering low friction will lock up while the other continues to turn. Aside from a slight lack of control, this does not produce a problem. It is when the locked-up wheel leaves the patch of ice that the trouble begins. At that instant, the brake-locked, zero-rpm wheel is being dragged over a high-friction surface by the momentum of the airplane.

Speed is a nemesis of tires; whenever safely possible, operate as slowly as you can. Excessive speed produces more internal heat and requires more braking. Touch down as slowly as practical for the conditions, do full-stall landings, then taxi slowly to the tiedowns. While taxiing, be particularly aware of turning. Tight turns are a significant cause of tread wear, especially pivoting about one wheel using the brakes. This creates a terrific sheer force, not only on the main gear tires, but on the nosewheel. Tight turns force the nosewheel to flex, scuffing the sidewall area and causing the tire to go out of balance. This is one of the primary causes of nosewheel shimmy and greatly reduces tire life.

Tight turns also put unnecessary stress on tire casing, beads and sidewalls. They produce flat spots that put the tire out of balance, causing it to thump, which in turn makes

Fig. 20-1. *Bald tires must be replaced.*

the airplane bounce while taxiing. Tight turns on gravel can cause a piece of stone to screw into the tire and puncture it. Often, because of the nature of the puncture, the flat tire isn't discovered until the next trip to the airport (FIG. 20-2). If a tight turn is impossible to avoid, allow the inner tire to make as large an arc as possible to minimize potential damage.

INFLATION PRESSURE—KEY TO TIRE LONGEVITY

Always keep tires inflated to the value specified in the pilot's operating handbook (POH) rather than the tire manufacturer's specifications. While this might sound contrary to the usual practice in aviation, the POH accurately reflects the actual tire loading for that specific airplane. The specifications from the manufacturer do not take into consideration the specific tire-loading value. In short, the airframe manufacturer knows the specific application—the tire manufacturer doesn't. The single most important preventive maintenance procedure the owner can do is to keep the tire pressure correct at all times. This will produce more landings per tire than anything else.

When new tires are installed, they typically are inflated without the aircraft weight acting upon them. Because aircraft weight deflects the tire, approximately a 4 percent

Fig. 20-2. *Pivoting turns on gravel lead to flat tires.*

increase in inflation pressure is required. New tires should be allowed to sit for 24 hours after installation so they can stretch and adjust to the rims. It is also important to check tire pressure daily during the first week because of probable air leakage.

Tire pressure should be checked with an accurate dial gauge prior to every flight (a preflight item that is seldom done by the average pilot); the heavier the airplane, the more critical this check becomes. It must be done on a cold tire (at least 4 to 5 hours since the last use), because a hot tire gives a higher reading. When going on a long cross-country flight, tires should be inflated for the coldest condition to be encountered. Ground temperature changes of 50 degrees Fahrenheit or more require greater inflation to compensate for the lower temperature. A good rule of thumb is every 5 degrees Fahrenheit temperature change yields a 1 percent change in tire pressure.

Improper inflation causes uneven tread wear. An under-inflated tire is subject to excessive wear on the shoulder. This is the worst possible condition, because it scars the sidewalls and shoulder as it rubs against the tire-rim flange. The result is faster heat buildup, and in extreme cases, tube tires can actually slip around the rim, shearing off the valve stem. Overinflation, on the other hand, causes excessive wear in the center of the tire, reduces traction and ground-handling ability, increases landing distances, and makes tire treads more vulnerable to cuts and nicks.

PREFLIGHT

During the preflight inspection, look carefully for cuts and nicks; they are not always easy to see. Most cuts and nicks can be avoided if the pilot uses caution and always watches the taxiway ahead. Foreign objects on runways and taxiways are a prime cause of cuts and nicks and should be reported immediately. If they are unavoidable, either get out of the airplane and remove them yourself, or call Unicom/ground control and ask for assistance. With the price of tires today and the safety factor involved, patience has never been a greater virtue.

Another common problem is potholes, large surface cracks, and similar drop-offs sometimes encountered between tiedown areas and taxiways. These should be avoided when possible, because they really can take a chunk out of a tire. If they are unavoidable, the only solution is to take them as slowly as possible. Be particularly careful of drop-offs; not only are they a problem for the tire, but if deep enough, you could also catch the prop.

The part of the tire that makes contact with the surface is called the *footprint*. A tire can be compared to a running shoe: both have tread gripping the ground, and the less tread, the less grip. The groove between treads allows water to pass under the tire without losing its grip on the surface. Earlier I mentioned skidding, where one tire locks up while the other continues to turn. This leads to blowouts, loss of control, and significantly increased landing rolls. There is yet another way to lose surface grip on a runway—hydroplaning.

HYDROPLANING

There are three types of hydroplaning: *viscous*, *reverted rubber* and *dynamic*. Viscous hydroplaning occurs on a smooth runway that has an extremely thin layer of water on it. It derives its name viscous from the fact that the water is actually in a semi-fluid state: ice changing into water. In this case, the tire never penetrates the water; it actually skis the entire time. The result is a total loss of control. It can happen at a speed significantly below what you normally would anticipate hydroplaning to occur—even at a fast taxi!

Reverted rubber hydroplaning is a bit more complicated. The necessary conditions are a wet surface runway and a skid in progress. When the tire locks up because of hard braking on a slick surface, the resultant friction generates heat, and the tire begins to smoke. Rubber debris collects under the tire, causing water to build up in front of and under the tire. The heat turns the water into steam and the tire actually rises up and floats on the steam. There is no runway contact, a severe loss of control, and, yes, it really can happen in light aircraft as well as heavies.

Dynamic hydroplaning, the third type, is what most pilots have encountered. The condition requires 1/10 inch or more of standing water on the runway. Excessive tread wear, and overinflation complicates the problem as tires alternately roll on the dry surface and ski on the wet. Virtually every pilot who has taken off or landed during heavy rain has experienced this phenomena, though they might not know it as hydroplaning. In the airplane, it feels as if the airplane is alternately sliding and jerking, typically from side to side.

There is a simple formula to determine at what speed an airplane will hydroplane: multiply 8.6 times the square root of the tire pressure in pounds per square inch. That gives you the lowest entry speed, however once hydroplaning has begun, it can continue at lower speeds! For instance, the Cessna 172Q nosewheel holds 45 psi and the main gear wheels hold 38 psi. That means the nosewheel will begin hydroplaning (if the right conditions exist) at 50 knots and the main gear at 46 knots. A Beech 58A Baron nosewheel has 55 psi, and the main gear has 52 psi; hydroplaning is at 56 and 54 knots respectively. Obviously, the best plan of attack is to try to touch down below those speeds if there is standing water or other conditions conducive to hydroplaning. Further, it is advisable not to use brakes until the aircraft has decelerated below the calculated hydroplaning speed; instead, you should continue aerodynamic braking until then.

Remember, hydroplaning can happen anytime you are at or above the critical speed, whether you are landing or taking off. Because the grooves between tire treads play an important role in getting water out from under the tire, it is important that they be checked on each preflight. The best method is to use a manufacturer-approved depth gauge, but at least check them visually.

PREVENTIVE MAINTENANCE

Often overlooked is the effect of contaminants on tires. It is not uncommon to see an airplane tied down with a tire sitting in a pool of oil or hydraulic fluid. All mineral-based fluids deteriorate rubber. Gasoline, oil, hydraulic fluid, grease, and tar are pure poison for tires. Any foreign substance on a tire should be cleaned off with soap and water. If that doesn't do the trick, take it off with gasoline, then remove the gasoline with soap and water.

Tires on airplanes that normally are tied down outside have an additional problem: weather checking. Harsh outdoor elements eventually lead to *checking* of tires (little hairline cracks). The tires essentially dry out and become somewhat brittle. Provided the plies are not visible, this should not cause any significant problems. Similarly, sunlight and static electricity in the air convert oxygen to ozone, which also attacks rubber and causes aging. This can be avoided by putting flexible, aluminum-coated covers over your tires

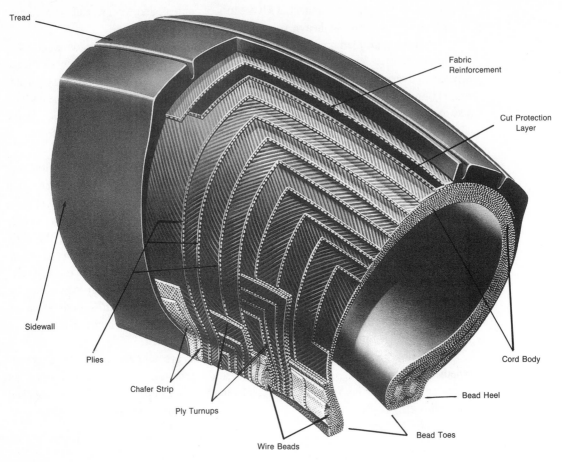

Tread

Fabric
Reinforcement

Cut Protection
Layer

Sidewall

Plies

Chafer Strip

Ply Turnups

Wire Beads

Cord Body

Bead Heel

Bead Toes

Fig. 20-3. *Aircraft tire cross-section.*

when the airplane is not in use. Incidentally, if you store your airplane in a hangar, don't breathe a sigh of relief too quickly. If there is an electric motor running inside the hangar, it also can interact with oxygen, create ozone and go to work on your tires.

Goodyear Aerospace, a leader in the tire industry, recommends replacing tires whenever the tread is worn to the base of any groove or to the minimum depth specified by the airframe manufacturer. If hydroplaning seems to be a persistent problem, tires with only $\frac{2}{32}$ to $\frac{3}{32}$ of an inch of groove depth should be removed. But remember, this is only a guide. There are many variables to tire replacement schedules.

Always remove a tire if there is any ply fabric exposed anywhere! Also, if a cut or crack is greater than 50 percent of a tread rib, if it goes to the base of a tread groove, or if it undercuts a tread rib, remove it (FIG. 20-3). Finally, bulges in any part of the tread, sidewall or bead area indicate separation and permanent damage to the tire and require removal.

180

21

Get Comfortable with Environmental Systems

Large turbine-engine aircraft, with their sophisticated environmental systems, are able to isolate occupants from outside weather. Naturally, the light airplane traveler does not have so many luxuries.

For one thing, the airlines literally have a trick up their sleeve—the airport jetways (boarding tunnels). Even the most sophisticated air conditioning and heating systems can't keep a cabin comfortable if it has a door open to the world. The jetway cleverly forms a sleeve that connects the airplane to the terminal, allowing maintenance of cabin temperature.

At terminals without jetways, a quick turnaround reduces the time cabin doors are open. Then too, airline cabins are long and have several bulkheads, minimizing the flow of air out the door. Understandably, the relatively small size of the average general aviation fuselage and the inadequacy of heating and cooling systems preclude the luxury of a stable cabin environment, unless you are able to load and unload passengers in a temperature-controlled hangar.

TYPES OF LIGHT-AIRCRAFT HEATERS

There are two types of heaters for light, general aviation airplanes: an *exhaust-manifold heater* and a *combustion heater*.

Exhaust Manifold Heater

The exhaust manifold heater, used exclusively in reciprocating, single-engine aircraft,

181

is the simplest. Working on a simple heat-transfer principle, a shroud is placed around the engine exhaust stack (FIG. 21-1). Fresh outside air is forced by ram pressure through the shroud and around the exhaust stack. The stack isolates the exhaust gas from the fresh air but allows the heat to transfer. To vary the temperature, the pilot controls a source of additional outside air that mixes with the heated air to cool it to the desired temperature.

The main advantages of the exhaust manifold heater are simplicity, low maintenance, and virtually no reduction in flight performance. It doesn't consume fuel, reduce engine power, or decrease airspeed by any detectable amount. The disadvantages, however, are obvious to anyone who has ever flown in a lightplane during winter. An exhaust manifold heater is a very ineffective system on the ground, because minimal ram air moves through the shroud. Not only does this mean the cabin tends to stay cold during ground operations, it also means poor windshield defrosting. Because windshield defrosting is accomplished by the rechanneling of cabin heat, it is common for pilots to taxi and take off with fogged windshields, peering through small circles smeared away with the back of their hands.

There is also a threat of carbon monoxide (CO) poisoning if the exhaust stack leaks inside the shroud. During preflight, it always is a good idea to check exhaust-stack seams

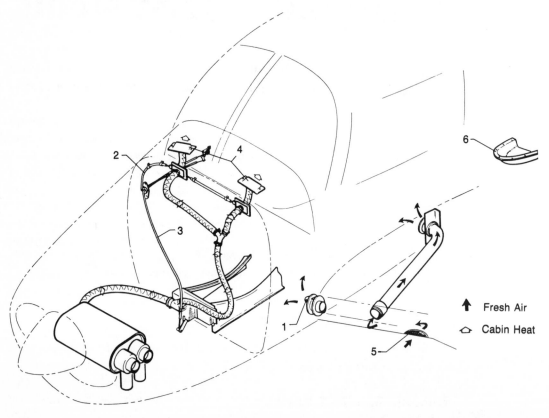

Fig. 21-1. *Exhaust manifold heater.*

to assure the welds are solid; any leakage within the engine compartment can cause the cabin to fill with CO. It is impossible to check the exhaust weld within the heater shroud, however, and it provides a direct route for CO to travel into the cabin. The best safety measure is to put an inexpensive carbon monoxide detector in every airplane. When exposed to CO, the colored pad turns from orange to black in 15 minutes or less (depending on the amount of carbon monoxide), alerting the pilot to a serious hazard.

Combustion Heater

If you think there ought to be a better way, you're right—there is. It's called a combustion heater, and most multiengine airplanes have them. But why don't singles? According to Bob Deeds, Customer Service Engineer for Janitrol Aero Division of Midland-Ross Corporation, Janitrol approached the major airframe manufacturers several years ago with the idea of installing combustion heaters in single-engine aircraft. The manufacturers balked at the idea because of expense, so at least for the time being, only multiengine pilots will have the luxury of instant heat.

Piper installs Janitrol combustion heaters exclusively in all of its twin-engine aircraft built after 1964. Janitrol shares the rest of the market with Stewart Warner's Southwind heater. The fundamentals of aircraft combustion heating haven't changed much over the years. According to the Aircraft Heating Digest (Volume 1, Number 1, published by Janitrol Aircraft in February, 1949), there are four main requirements: fuel for combustion, air for combustion, ignition to start combustion, and air to carry away the heat produced by combustion. That was true for the DC-3, and it's still true for modern aircraft.

Heat is produced by burning a fuel/air mixture in a heater combustion chamber. This is somewhat of a mixed blessing, because while it conveniently uses fuel drawn from the aircraft fuel tanks, it also reduces the aircraft's range when the heater is in use. Nonetheless, when you get into a cabin that is below freezing and can have near instantaneous heat without even starting an engine, a little less range—at least to this northerner—doesn't seem so bad.

The Janitrol heater uses a spray nozzle to send regulated, atomized fuel/air mixture into the heater combustion chamber. There, a high voltage spark plug powered by the aircraft's electrical system provides continuous ignition. Because aircraft attitude and altitude are always subject to change—sometimes rapidly—Janitrol uses what it calls the ''whirling flame'' principle. The fuel/air mixture enters the combustion chamber tangent to the chamber's surface (FIG. 21-2). This forces the airflow to spin and mix with itself, causing a stable, continuous flame pattern. The burning gases flow the length of the combustion tube, double back over the outside of the chamber, go through a crossover passage to an outer radiating area, travel down the length of the heater one more time, and finally exit through the exhaust.

The cabin ventilation is ducted separately between the combustion air chambers. Though the two airflows never mix (to do so would lead to CO poisoning), the ventilating

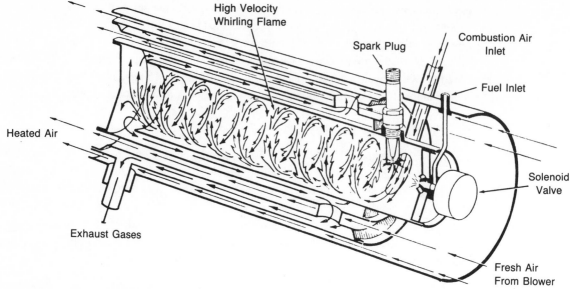

High Velocity
Whirling Flame

Spark Plug

Combustion Air
Inlet

Fuel Inlet

Heated Air

Solenoid
Valve

Exhaust Gases

Fresh Air
From Blower

Fig. 21-2. *Janitrol heater cutaway.*

air does contact several surfaces heated by the combustion air, causing heat transfer. Several other components round out the system. An electric fuel pump is necessary, though some aircraft actually use the engine-driven fuel pump if the fuel pressure output is correct. There must be a ventilating air blower, which also doubles as a cooling fan when the heater is turned off, and a separate combustion air blower. Temperature control is maintained by a duct switch that senses heat output and compares it to the selected temperature. And of course there are the requisite controls and lights that indicate the operational status of the heater.

PREFLIGHT

System preflight should include checking for either blockage or damage to both the ventilation and combustion air inlets, the heater-exhaust outlet and the heater fuel drain. Additionally, the area around the heater exhaust tube should be checked routinely for soot accumulation; this indicates an excessively rich fuel/air mixture that is caused by incorrect fuel pressure, a blocked combustion-air inlet line, an inoperative or failed combustion blower, or a clogged fuel nozzle.

In addition to the visual preflight, an operational check should be done. First, turn the heater master switch on and assure the ventilation and combustion blowers work. The heater-failure light should also illuminate, indicating that the system is activated, but there is no combustion. With the master switch still on, check for excessive current draw or any unusual vibrations or noises. Then perform the operational check as outlined in your pilot's operating handbook (POH).

TROUBLESHOOTING

Six basic operational problems can occur that require troubleshooting on the pilot's part: heater fails to light; ventilating air blower fails to run; combustion air blower fails to run; heater fires but doesn't burn steadily; heater starts, then goes out; and heater fails to shut off. If the heater fails to light, the first consideration is to double check the POH to make sure you are doing the correct procedure. Beyond that, insufficient electrical power such as a dead battery or insufficient fuel should be suspected. Some mechanical problems are beyond the control of the pilot, such as restricted fuel nozzle or inoperative fuel pump. Failure of the ventilating blower means you probably forgot to turn the heater master switch on. Otherwise, it's a mechanic's job, as is failure of the fully automatic combustion air blower. When the heater fires up but doesn't burn steadily, the culprit is probably fuel related, for example insufficient fuel or contamination by ice or water. Other mechanical problems, such as a fouled spark plug, can produce the same results. If the heater just goes out, mechanical problems could be the cause. More likely, it is either fuel or electrical starvation; check the fuel supply and master switch. Finally, if the heater fails to shut off during shutdown, it is a mechanic's problem, such as a defective heater switch or stuck fuel solenoid valve.

The crux of the matter, according to Deeds, is proper maintenance. Preflight and preventive and periodic maintenance are the keys to efficient and safe operation. Your new Janitrol heater is certified to run 500 hours (or 24 months, whichever comes first) before periodic maintenance is required. After that, another inspection is due every 200 hours (or 24 months, whichever comes first). Performed by an A&P, the preventive maintenance is a thorough inspection of the entire unit, including a pressure check of the combustion chamber. Incidentally, it is important to note that the "hours" referred to are actual heater-operation hours. While some aircraft have a heater hour-meter that records operating time, many do not, so 1 heater hour is computed to be the equivalent of 2 flight hours.

AIR CONDITIONING

It wasn't all that long ago that the sign, "It's Cool Inside" was emblazoned across a theater marquee and was enough to attract crowds, regardless of the picture. Air conditioning spread to restaurants, to other public places, and finally to homes. Previously inured to heat, consumers quickly began to expect to be kept cool indoors; portable units found their way into cars and finally commercial aircraft. If asked, the average pilot would probably tell you it isn't practical to air condition small aircraft. We tend to be most concerned about protecting the pilot from extreme cold, forgetting that excessive heat can also be a problem.

Studies conducted by both the U.S. Air Force and the U.S. Army show an airplane with a stable, comfortable cabin temperature is a safer flying environment. With 30-minute waits on the ground at some of the larger airports and outside air temperatures of 80 to

90 degrees Fahrenheit, it's no wonder the inside of an airplane can exceed 100 degrees. How safe can a pilot be after sitting in a 100-degree cabin for 30 minutes prior to takeoff? Few business executives are going to sit in that kind of heat.

However, there are penalties to be paid for air conditioning, to be sure. There is an increase in aircraft empty weight due to the compressor and other required equipment. This translates into fewer bags, reduced fuel, or fewer passengers. Just the operation of the system causes a reduction in available engine horsepower. The Cessna 210N operating handbook states there is a 1-knot TAS cruise reduction when air conditioning is simply installed on the aircraft, and an additional 1- to 2-knot TAS cruise reduction when the compressor actually is operating!

Two types of heat affect airplanes: aerodynamic and solar. Aerodynamic, also known as *adiabatic skin temperature*, is the result of free-stream kinetic energy being converted to thermal energy when the free stream air is slowed to zero at the surface of the airplane. The faster the airplane moves through the air, the greater the heat buildup, because skin temperature is a function of free-stream temperature and Mach number. For instance, at Mach 2.0, the fuselage temperature would be approximately 260 degrees Fahrenheit; at Mach 5.0, it would be about 1550. This obviously is a problem for large aircraft, not singles or light twins. For the slower aircraft, the basic problem is the sun and little or no ventilation to carry off cabin heat. The automotive type air conditioner fits nicely into this type of airplane.

Fundamentally, air conditioning is simple physics; the rapid expansion of fluid causes a drop in temperature. There are two basic types of air conditioning units; *air cycle machines* (ACM) and *vapor cycle systems*. Large aircraft ACMs bleed compressed air from the turbine engine and allow it to expand, causing cooling. With the vapor-cycle system used in light aircraft, a pressurized liquid refrigerant evaporates, causing a temperature reduction. This liquid refrigerant, called freon, is usually F-21 (dichloromonofluoroethane) or F-12 (dichlorodifluoromethane). These two chemicals (which are not interchangeable) are used because they are nonflammable, nontoxic, and do not cause irritation.

A sample system consists of a compressor, a condenser, a receiver/dryer, an expansion valve, an evaporator, and blowers (FIG. 21-3). The compressor is driven by the engine through a belt-and-pulley system, and a clutch disconnects it when cooling is not required. The compressor discharges high pressure, higher temperature refrigerant vapor.

The refrigerant then passes through copper coils surrounded by cooling fins, the condenser, to maximize refrigerant heat transfer to outside air. The condenser hangs under the fuselage in most aircraft and retracts into it when the system is inoperative. Because of the excessive amount of drag the condenser causes, most aircraft have a throttle interlock switch that automatically retracts it when full power is applied while simultaneously disengaging the compressor clutch from the engine. After leaving the condenser, the liquid refrigerant goes to the receiver/dryer.

The receiver/dryer functions as the system's reservoir. It contains a desiccant (typically silica gel) which absorbs moisture. (A single drop of water can freeze, lodge in the expansion

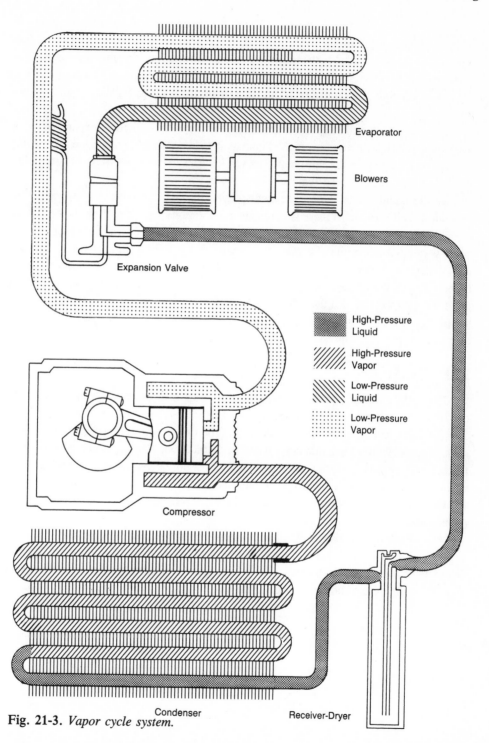

High-Pressure
Liquid

High-Pressure
Vapor

Low-Pressure
Liquid

Low-Pressure
Vapor

Evaporator

Blowers

Expansion Valve

Compressor

Condenser

Receiver-Dryer

Fig. 21-3. *Vapor cycle system.*

valve and completely stop the system! There also is another problem when water and refrigerant mix: they form highly corrosive hydrochloric acid that eats up the system from within.) A filter is installed in the receiver/dryer to prevent blocking the expansion valve. There also is a sight glass in the receiver/dryer, similar to the one in automobiles. With the system running, the sight glass should appear perfectly clear; bubbles mean low fluid level.

The refrigerant, still a high pressure liquid, continues to the thermal expansion valve. This valve reduces the pressure of the liquid as it passes through, meters the refrigerant, and maintains high pressure upstream. The valve varies refrigerant discharge, depending on the amount of heat to be removed from the cabin.

As the liquid leaves the valve, it is sprayed through the coils of the evaporator, expanding as it goes, until there is complete evaporation of the liquid by the end of the coils. The evaporator is effectively the low pressure equivalent of the condenser unit and consists of parallel circuits of copper tubing with fins. As the warm cabin air circulates around the evaporator fins with the help of the blower, the heat of the cabin air transfers to the refrigerant as it passes through the evaporator, and cooler air remains to continue on into the cabin. The heat the refrigerant picked up raises the refrigerant's temperature to boiling, causing it to change state from liquid to vapor. The low pressure vapor cools as it returns to the compressor, where it is pressurized to begin the process again.

PREFLIGHT

Preflighting the air conditioner consists of a visual inspection of the compressor, the drive belt and pulley system, all hoses, the condenser inlet, and condensation outlet drain. The emphasis should be on system integrity. When the system is turned on initially, it should operate within 1 to 2 minutes; otherwise shut it down. One indication of insufficient freon is little or no cooling air and a buildup of frost on the evaporator. A hissing sound in the evaporator is yet another indication. As with all systems, it is important to read and follow the POH carefully with respect to preflight, operation, preventive, and periodic maintenance.

As adaptable as humans have proven themselves to be over the centuries, they still have a very limited temperature range. Within that range, there is an even smaller one that dictates comfort and efficiency. Therefore, there is truth in the belief that properly maintained environmental systems promote safe, comfortable flying.

22

Know Your Aircraft Pressurization System

The main reason for pressurizing an aircraft is flexibility. Being able to select a higher altitude can give you the option of a smoother ride, shorten your flying time, and/or provide an alternative to flying in severe weather or icing conditions. A pressurized aircraft can provide a comfortable cabin environment at significantly higher altitudes than one that is unpressurized.

Cabin pressure is maintained by "packing" air into the cabin at a fairly constant flow and then controlling the flow of air out of the cabin. Because there already is a constant flow of air out of the cabin through cracks, door, and window assemblies and other leakage points, it is far easier to pump in more air than is necessary and maintain the desired cabin pressure by regulating the opening of an outflow valve. This constant airflow also keeps cabin air fresh.

According to Bob Deeds, technical representative of the Product Support Division of Janitrol Aero, his company's pressurization system for the Cessna P210 has four basic modes of operation: *unpressurized*, *isobaric*, *differential*, and *negative relief*. The unpressurized mode is in effect any time the aircraft is at a lower altitude than the cabin altitude requested by the pilot; this is common during takeoff, climb, descent, and landing. The isobaric mode begins when the aircraft climbs through the selected cabin altitude, which can range from below sea level to 10,000 feet. In the P210, the pilot selects the desired cabin altitude on the manual controller prior to takeoff; no other input is required through takeoff, climb, and level-off. If a change of aircraft cruise altitude is required, the pilot slowly adjusts the controller to preclude abrupt cabin altitude changes which are felt by the passengers as pressure "bumps."

The manual controller has two altitude scales (FIG. 22-1). The outer scale indicates cabin altitude; the inner scale indicates the corresponding aircraft altitude at the maximum operating cabin pressure differential. (Cabin pressure differential is the ratio between inside and outside air pressures.) These numbers on the controller face must be multiplied by 1000 to determine the appropriate altitude in feet. The pilot turns the cabin rate control knob to adjust the rate at which the cabin pressure ''climbs'' or ''descends'' to the altitude he has set on the manual controller. The differential pressure mode goes into operation whenever the maximum cabin-to-ambient pressure differential is reached. Because differential pressure is a measure of internal stress on the fuselage skin, if it were to become too great, structural damage to the fuselage might occur.

The transition from the isobaric mode to the differential control is automatic. The operating differential normally is maintained by the outflow valve with the safety valve acting as a backup, allowing a pressure differential only slightly higher than what is regulated by the outflow valve.

The maximum pressure differential value varies from aircraft to aircraft, depending on system and structural limitations and the type of operation for which the aircraft is designed. An aircraft such as the Beech Baron 58P, which has a maximum cabin differential pressure of 3.65 pounds per square inch (psi), is limited to a difference of 3.65 psi between the ambient air pressure and the cabin air pressure. On the ground, there is a 1:1 cabin-pressure-to-ambient-air-pressure ratio, but as the Baron 58P climbs, that ratio changes until there is a 3.65 psi pressure difference. The 58P is capable of maintaining sea level pressure up to approximately 7000 feet (TABLE 22-1), where the difference between the

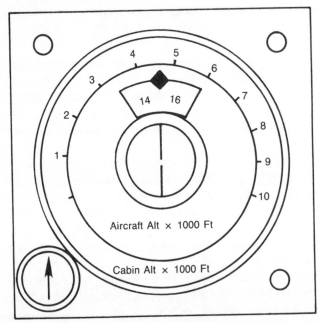

Fig. 22-1. *Cessna 340 controller face.*

Table 22-1. Standard Atmospheric Pressure

Altitude (feet)	Pressure (psi)	Altitude (feet)	Pressure (psi)
Sea Level	14.7	18,000	7.3
1,000	14.2	19,000	7.0
2,000	13.7	20,000	6.8
3,000	13.2	21,000	6.5
4,000	12.7	22,000	6.2
5,000	12.2	23,000	5.9
6,000	11.8	24,000	5.7
7,000	11.3	25,000	5.5
8,000	10.9	26,000	5.2
9,000	10.5	27,000	5.0
10,000	10.1	28,000	4.8
11,000	9.7	29,000	4.6
12,000	9.3	30,000	4.4
13,000	9.0	35,000	3.6
14,000	8.6	40,000	2.7
15,000	8.3	50,000	1.7
16,000	8.0		
17,000	7.6		

sea level pressure and the 7000-foot atmospheric pressure is 17.7 psi − 11.3 psi = 3.4 psi, nearly the maximum pressure differential. At 21,000 feet and maximum cabin differential pressure, the cabin altitude would be approximately 10,000 feet.

Another factor that determines the maximum possible cabin pressure is the type of pressurization system used. The higher the aircraft is designed to operate, the greater the maximum differential needed, and the stronger the compressor output capacity required. Turbine engines can maintain high pressure airflow into the cabin up to very high altitudes by using air from the compressor bleed-air section of the engine. Aircraft with reciprocating engines use air from the compressor section of the turbocharger (FIG. 22-2). In light twin-engine aircraft, powerplant failure, or even a significant, intentional power reduction, can cause the cabin altitude to rise when there is a high cabin pressure differential. This can occur because the turbocharger—which is the source of the pressurizing air—is powered by the engine. Power reductions in single-engine aircraft have the same effect, so descents should be initiated far enough in advance so the power will not have to be cut back. For the same reason, pilots should be careful not to run a fuel tank dry in a pressurized aircraft; depending on the amount of uncontrolled cabin leakage, cabin altitude could rise faster than you can switch tanks and get the engine running again.

When air is compressed, it increases in temperature. Turbine-engine bleed air is so hot it always requires cooling before entering the cabin, even if warm air is desired. Larger turbine-powered aircraft run the pressurized bleed air through air-conditioning packs prior

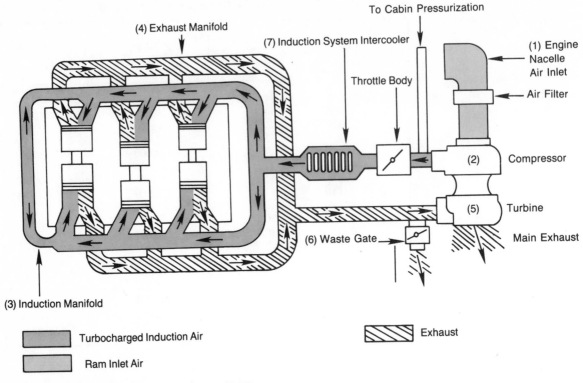

Fig. 22-2. *Cessna 421 turbo system schematic.*

to cabin entry. Air from a reciprocating-engine turbocharger may require cooling only on warm days when the aircraft is flying at lower altitudes. This typically is accomplished by routing the air through a heat exchanger where the pressurized air ducting is cooled by ambient ram air. At very cold ambient temperatures when considerable heat is required in the cabin, the pressurized air might not be warm enough, and a cabin heater will be required.

The outflow valve, which vents the cabin to the outside air, has three main functions: *negative pressure relief*, *isobaric control*, and *differential control*. Negative pressure relief is automatic, so the aircraft is never subjected to an outside air pressure greater than cabin pressure; higher pressure air always can flow freely through the outflow valve into the aircraft.

Isobaric control typically maintains cabin pressure within ±0.05 psi from what the pilot selects on the manual controller. If the cabin pressure exceeds that selected on the controller, the outflow valve increases to compensate. As the pressure falls below, the opening decreases slightly.

Differential control is preset by the factory. The isobaric pressure requested by the pilot will be maintained until cabin pressure reaches the maximum pressure differential.

Then, the differential control overrides the isobaric mode so the cabin altitude will vary directly with aircraft altitude.

If the outflow valve were to stick closed, excessive cabin pressure would build up quickly. To prevent over-pressurization resulting from a stuck outflow valve, the system has a safety valve that functions to relieve negative pressure, to provide backup differential control, and to act as a solenoid-operated cabin-pressure dump mechanism that can be operated from the flight deck or by a gear squat switch.

RAPID DEPRESSURIZATION

One of the most misunderstood aspects of cabin pressurization is rapid depressurization. Often inappropriately referred to as "explosive decompression," movies have depicted scenes where a single bullet shot through the fuselage has sent people flying about the cabin. If you can have a fairly large outflow valve open to the atmosphere, what difference would a small bullet hole make? As long as the bullet hole remains small, the outflow valve can compensate instantly.

A pilot can depressurize the cabin intentionally if the pressurization system begins to pump contaminated air into the cabin. Also, if the pilot discovers a window is cracking, depressurization will relieve the pressure differential and take stress off the window.

There are two ways to depressurize a cabin intentionally. The gentlest would be to increase the cabin altitude slowly with the manual controller until there no longer is a pressure differential; this would be an appropriate method in the case of a cracked window, or the presence of limited smoke or fumes. In a life-threatening situation, such as a cabin fire or dense smoke, the pilot could elect to activate the depressurization switch. This reduces the pressure differential to zero rapidly, but not instantly—the outflow and safety valves are not that large!

One of the frequent causes of rapid decompression is improper closure of a door. Unlike larger aircraft with plug doors that are pushed into the fuselage by cabin pressure, light aircraft doors generally open outward, away from the fuselage. The manufacturer has to devise a locking mechanism that is easy to use but strong enough to ensure the door will not open in flight. If someone incorrectly locks the door, it could blow out when the pressure differential increases. The potential for this problem, which is almost always caused by human error, is significant enough in the King Air 200 that when I worked for FlightSafety, we used to instruct all crews never to permit passengers to lock the cabin door—it was to be locked only by a crewmember. This is good advice for all pressurized aircraft.

Even if a door or window did burst open in flight, it isn't very likely that passengers of light aircraft are going to be sucked out of their seats and pulled through the door. An analogy would be if you filled the airplane with water and punched a hole in the fuselage; the bigger the hole, the more the water would tend to pull you toward it.

There should be concern for people who have ear or sinus trouble, however, because instantaneous pressure changes can be painful. Cold, the sudden feeling of fear, and lack of oxygen present a far more serious problem than the rapid decompression itself. It feels as if someone has stepped on your chest, the air suddenly expands in your lungs, forcing its way out through your nose and mouth. You couldn't hold your breath if you tried, but it would never occur to you to try. The cabin goes IFR for a short time and it can become incredibly cold. Panic could be a problem for passengers with heart trouble.

Losing cabin pressure when flying above 10,000 feet probably will warrant immediate descent. Even with supplemental oxygen available, the cold can be deadly. The best course of action for the pilot is to put on an oxygen mask immediately, pause for just a few seconds to shake off the fear, then check the passengers to make sure they are on oxygen. Then the pilot can assess the situation and determine the best course of action based on cabin temperature, oxygen available, structural condition of the aircraft, weather, distance to the airport, wind conditions at a lower altitude, and the condition of the passengers. It is important that the pilot be familiar with applicable emergency procedures in the pilot's operating handbook. He or she should be aware that a lower altitude frequently means increased turbulence, so descent should not be made at V_{NE}.

Another concern about rapid descent is existing structural damage, because the airframe might not hold up under high speed descent. Also remember that as you descend, air density increases and so will the indicated airspeed for a given deck angle.

There are other concerns related to cabin depressurization. If while sitting in the freezing cold contemplating a course of action sounds like a great time for a cup of hot coffee, remember the thermos was sealed at ground pressure, so it is a potential bomb. Don't forget to squawk 7700 on the transponder. If you are in instrument conditions and it becomes necessary to use alternate instrument air, be aware that a large hole in the fuselage can cause the cabin pressure to be lower than ambient due to a venturi effect. Perhaps most often overlooked is a passenger briefing before the flight—a little knowledge can go a long way, especially concerning the use of oxygen and effects of smoking.

PREFLIGHT AND OPERATIONAL CONSIDERATIONS

During the preflight, make sure the door is properly sealed and the dump switch is off. After engine start, to assure the system will work while still on the ground, set the aircraft altitude controller to 500 feet below field elevation. Then, pull the landing gear circuit breaker and increase the rate controller; the system should begin to pressurize the cabin because you have overridden the gear squat switch and tricked it into "thinking" it was flying above the selected altitude. However, you should never take off in a pressurized condition, because the aircraft is not designed for it.

In preparation for takeoff in the Cessna 340, which uses the Garrett AiResearch system, the procedure is somewhat different than in the Cessna 210. AiResearch instructs pilots to select 500 feet above field elevation on the cabin altitude selector and set the cabin rate control knob to the 12 o'clock position. Then start the engines and check for airflow into the cabin to assure it will pressurize after takeoff. There are two reasons for doing this: First, it prevents the pressure "bump" sometimes felt on takeoff as a result of both the safety and outflow valves closing simultaneously. The safety valve that closes when the gear retracts is controlled by the squat switch. The outflow valve closes when the cabin reaches the altitude you have requested. If the controller is set to field elevation, both can slam shut simultaneously on takeoff. With the controller set to 500 feet above field elevation, the outflow valve will close long after the safety valve and the passengers will experience a smoother transition. The second reason is if you set the selector to cruise altitude, the system will "prerate," meaning it will think the aircraft is climbing long before it actually does. This will delay normal cabin pressurization longer than necessary and could prove uncomfortable for some passengers.

Once the climb is established and you have passed through 500 feet AGL, reset the aircraft altitude selector to 1000 feet above cruise altitude. As the aircraft climbs, the cabin altitude takes care of itself. The reason for setting 1000 feet above cruise altitude is, again, passenger comfort. If the controller is set to cruise altitude, the outflow valve will open and close continuously as the cabin pressure makes small fluctuations between too low and too high. With the controller set for an altitude above the actual aircraft altitude, the cabin will never reach the programmed pressure, so the outflow valve will remain at least slightly opened; the result is no bumps. There are no additional requirements for cruise condition. If it is necessary to change altitude, simply select the new altitude plus 1000 feet and climb or descend.

During descent for landing, set the aircraft altitude selector to approximately 500 feet above field elevation and adjust the cabin rate-of-change to maintain a comfortable cabin rate of descent. It is a good idea not to descend at a rate that will allow the aircraft to catch up to the cabin altitude—otherwise the cabin will depressurize. On the other hand, it you select the field elevation for the cabin altitude, when the gear touches down on the runway, the cabin will dump, causing some passenger discomfort. With 500 feet above airport elevation selected, the cabin will depressurize comfortably shortly before landing.

TROUBLESHOOTING

According to Dick Weatherly, customer service engineer for Garrett AiResearch Manufacturing company, pressurization systems are basically reliable. AiResearch recommends a few troubleshooting tricks:

✈ If the cabin follows the aircraft's altitude and rate-of-change shown on the flight instruments, there may be several causes. The first thing to check is the dump switch.

✈ If it is off, the logical choice would be a problem in the landing gear solenoid valve which is responsible for keeping the safety valve open during ground operations.

Try cycling the gear to see if that helps, then try opening the landing gear circuit breaker to bypass the system.

✈ If the down rate is faster than the up rate but everything else works normally, just make the necessary adjustment manually and have the controller checked out at your next opportunity. The problem probably is a minor leak in the tubing or controller.

✈ Should the cabin rate exceed the selected rate value during the aircraft's climb to cruise altitude, increase the rate selection or decrease the aircraft's rate of climb. The obvious answer is that the aircraft is climbing faster than the controller and it is at the maximum differential, however it also could be a controller malfunction.

✈ If the cabin altitude exceeds what you have selected, one of several problems could exist: there could be a loss of pressurizing airflow for some reason; the aircraft altitude might have exceeded the positive differential pressure value; or there could be an internal malfunction of the controller, the outflow valve, or safety valve. Your only choices would be either to adjust to a higher cabin selection, if possible, or to reduce the aircraft altitude. Other problems that may arise usually are beyond the pilot's ability to correct in flight.

The light aircraft pressurization system is, for the most part, both safe and reliable. It has been said that its bark is far worse than its bite, but a thorough understanding of the system, as well as a good passenger briefing, will go a long way toward keeping it that way.

23

How to Troubleshoot Deicing Systems

Ice is the nemesis of every IFR pilot. Virtually all aspects of it are negative: it's capricious; it's fickle; should you climb, descend, turn back? The stakes are high, particularly for those who fly IFR in aircraft not certified for flight into known icing. A chance encounter with icing can produce drastic aerodynamic changes, propeller imbalance and vibration, increased drag, increased weight, and reduced airspeed. To fight back, general aviation pilots can arm their aircraft with deicing equipment.

Simple as it may be, some folks don't understand the fundamental difference between deicing and anti-icing equipment. *Deicing,* as the name implies, removes accumulated ice from the leading edges of wings, horizontal stabilizers, vertical stabilizers and propellers. Yes, props have *de*icing, not anti-icing equipment!

Anti-icing equipment is used where no amount of ice can be tolerated. This equipment requires a significant current draw and, except for the relatively small pitot tube heat, is seldom found in light aircraft. A common location of anti-icing is at turbine-engine inlets, where a chunk of ice could produce disastrous results.

The first working airfoil deicer was designed in 1929 by William C. Geer in conjunction with the Guggenheim Safety Foundation and the National Advisory Committee for Aeronautics. His design was implemented by BF Goodrich in 1932, and the first pneumatic deicers to be installed on a commercial aircraft were on a 1930 Northrop Alpha mail plane. A close relative of those early pneumatic deicers is still used on the leading edges of airfoils. Since then, however, electrothermal deicers have been added to propellers.

ELECTROTHERMAL PROPELLER DEICING

The colder the outside air temperature, the greater the tendency for ice to adhere to a surface. Fundamentally, *electrothermal propeller deicing* is a simple matter of converting electrical energy into heat and transferring the heat to the prop. That takes a fair amount of electrical power, and when you are flying IFR in icing conditions, you don't have a lot of power to spare. To solve that problem, the job is broken into two elements: outboard and inboard. In a single-engine airplane, first the outboard element heats up, then the inboard. This cycling between outboard and inboard continues as long as the propeller deice switch is on. With a 14-volt system, you can anticipate a 20- to 13-amp draw with a two-blade prop and 30 to 34 amps with a three-blade one, as compared to 8 to 12 and 14 to 18 amps respectively for a 28-volt system.

When deicing, there is more at work than just heat. Centrifugal force is pulling constantly, especially on the outboard section. Here, a little ice buildup is actually desired, because by increasing the mass of the ice, it is more likely the ice will all come off at once. The pause in outboard heating while the inboard element is activated allows some buildup. Then, as the outboard heat turns on again, the adhesion of the ice reduces, centrifugal force tugs away, and the ice flies off into the blast of air. There may be a horrendous *whump* as the ice slams against the side of the fuselage.

The dual-element system, though not the only type, is still the most common. In the single-engine airplane, regardless of the number of propeller blades, there are two independent circuits (FIG. 23-1). All outboard elements simultaneously heat for 34 seconds, then all inboard elements for 34 seconds. If all outboard elements didn't heat simultaneously, there would be a strong tendency for rotational imbalance to occur as one prop outboard section shed ice and the other didn't. BF Goodrich now offers the HotProp. This single, graduated heat element solves the problem of excessive electrical load by providing high heat at the inboard end and low heat at the outboard end. The HotProp is still two-cycle: 34 seconds of heat over the entire length of all blades and 34 seconds of no heat.

Twin-engine aircraft compound the power demand by adding another set of prop blades. Dual element systems are still the most common, but here the cycling is different (FIG. 23-2). The sequence of these 34-second cycles is as follows: right outboard, right inboard, left outboard, left inboard. Rotational balance is assured the same way as with the single-engine aircraft system. It is important to note that the timer in the twin-engine aircraft system does not reset to a "home" position when deactivated, so you never know which elements will heat up first, but the sequence will always be the same.

An electrothermal propeller deice system can operate on either a 14- or 28-Vdc system. The actual deicer elements are hand-wound conductive wiring embedded in a thin fabric and rubber sheet that is bonded to the first third of the leading edge of the propeller blades. The principle is simple, but putting the principle into practice is what causes trouble. The problem is how do you get electrical power out to a rotating propeller? The solution—and weak link—is a slip-ring and brush-block assembly.

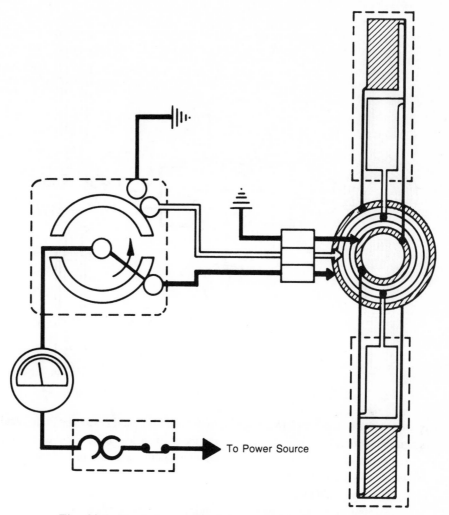

Fig. 23-1. *Dual-element deicing system (single-engine aircraft).*

This copper slip-ring and brush-block electrical distribution center transfers electric current to the rotating propeller deicers from spring-loaded carbon brushes. The brushes, fixed to the airframe, maintain constant contact with the slip ring as it rotates with the propeller. Potential problems range from uneven wear to loss of conductivity as a result of oil accidentally covering the slip ring. A common pilot-induced problem results if the system is turned on when the engine is not running; the stationary slip ring gets burned by the brush.

A timer continuously cycles the elements at 34-second intervals. Most systems also have an ammeter that indicates the operating current. Other than a momentary fluctuation

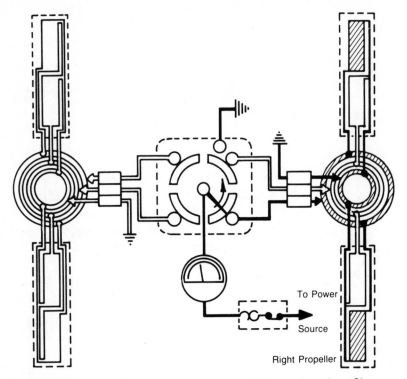

Fig. 23-2. *Dual-element deicing system (twin-engine aircraft).*

as the timer cycles to another element, the ammeter needle should always be in the shaded area when the system is activated. Prior to engine start, a low battery could cause the needle to be slightly below the shaded area. Rounding out the system is the circuit breaker, on/off switch, and the wiring harness.

Propeller deice should be checked prior to every flight during winter months and whenever you are going to fly into known icing conditions, if the airplane is certified for it. The rest of the time, you should preflight check it at least once a week. The prudent pilot will follow what the aircraft's pilot's operating handbook (POH) says about preflighting propeller deicers. While there are differences between aircraft, here are some general items to check:

✈ Visually inspect the rubber elements carefully. Look for wrinkles, debonding, rips, impact damage, cracks, or erosion of the rubber surface. Several leading-edge tapes designed to reduce erosion are available—DON'T USE THEM! BF Goodrich has conducted tests on several brands of these tapes and found they have insufficient thermal conductivity and result in a lower surface temperature than may be required to shed ice!

☜ After a good visual inspection, activate the system and put your hands on the elements. You should feel them get warm in a few seconds; if not, there is a problem. In either case, turn off the system to prevent the brushes from burning the slip ring. In fact, prolonged use also causes significant battery drain and the elements can do serious damage to non-metal propeller blades.

☜ Finally, conduct an operational check. With the engine running, turn on the deicing system and observe the ammeter. Every 34 seconds, the needle should deflect momentarily, indicating the timer is cycling and putting out power. It is important to understand that this doesn't necessarily mean the elements are getting the power. It only means that the timer is turning it on. With all of that slapping, scraping, rotating, and vibrating going on, there is more than just a chance things won't work exactly right. To head off trouble, BF Goodrich requires both a 50-hour and a 100-hour inspection by an A&P mechanic, and it's the cheapest insurance you can get.

The 50-hour inspection determines whether all of the current is actually getting to all of the deicing elements at the right time. The A&P will inspect the wiring harness carefully, measure current flow, assure proper sequence, and check the elements for hot spots that indicate an area of local trouble. The 100-hour inspection duplicates the 50-hour and adds a thorough check and cleaning of the slip ring and carbon brushes.

Troubleshooting the deicing system is a relatively simple task—you either get the *whump* or you don't. There is very little you can do once airborne. The ammeter is the key to virtually everything. If it reads zero, check the circuit breaker. On the ground, with the engine not running, it could be the battery master. If the ammeter reads *normal* during part of the cycle and *zero* during another part, you've got trouble. With partial deice capability, rotational imbalance is likely to occur that could be severe enough to cause structural failure of the prop. Deactivate the system. A *normal* ammeter reading during part of the cycle and *low* current during the other part probably means that inner and outer elements are heating simultaneously. That produces a hefty current draw, but probably nothing more serious. If the ammeter always reads low, you have low system voltage—an indication of generator or voltage regulator problems. A constant high reading indicates a deicer power lead is shorted to ground, warranting a system shutdown. If the ammeter flicks more frequently than every 34-seconds, there probably is a loose connection; less frequently indicates an inoperative timer and consequently incomplete deicing. The potential problem, again, is rotational imbalance.

If the propeller isn't shedding ice at all, there could be a short in the wiring harness, worn out carbon brushes, even gas or oil on the slip ring and brushes. If there is oil on the slip ring, have your prop seal checked—it could be worn out. If radio static occurs only when propeller deice is turned on, it is probably the result of arcing brushes, loose connections, or a wiring harness that's too close to the radio equipment. If you can't stand the popping for another flight, turn this aeronautical Gordian knot over to your A&P and make sure you set a maximum dollar amount on unraveling it!

PNEUMATIC DEICING SYSTEMS

If ice will adhere readily to a propeller, it will collect massively on a wing or horizontal stabilizer. Because the wing is so large, electrothermal heating really isn't practical. While more sophisticated aircraft have weeping wings, hot wings and other expensive equipment, most of us live with pneumatic deicing systems, what we affectionately call "boots."

Here, the principle and practice are relatively simple. As ice accumulates on the leading edge, you mechanically expand it and break the ice loose. This system consists of inflatable rubber deice boots, a pneumatic system to inflate them, timer and relay switches for inflation sequencing, an on/off switch, and a pressure gauge or indicator (FIG. 23-3). The heart of the pneumatic system is the pump: the pressure side inflates the boots; the vacuum side deflates and holds them down. Other components include tubing, which seems to run endlessly through the wings and fuselage; flow control valves to channel airflow to the boots; regulators to control both pressure and vacuum level; and pressure-relief valves.

The boots are fabric-reinforced rubber sheeting, bonded and stitched so as to produce parallel, inflatable tubes which may run along the span or the chord. All rows may inflate simultaneously or they may inflate alternately, but all systems are essentially the same. If you are not experienced at flying in icing conditions, be warned that dual instruction is necessary before tackling it alone; it is not as easy as turning on the boots and autopilot. You need to let about ½ inch of ice build up before activating the boots or it might not break away. Also, inflating the boots too frequently causes ice to build up around the extended boots. How frequent is too frequent? Now we're back to that dual I mentioned.

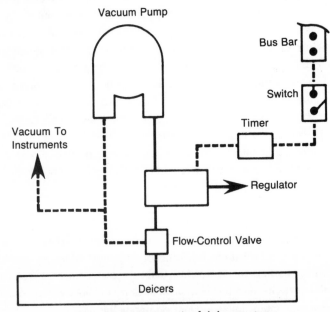

Fig. 23-3. *Pneumatic deicing systems.*

Preflight the boots by visually inspecting them for cracks, rips, tears, bubbles, holes, or separation from the airframe. You are observing the general condition of the rubber. During the cockpit check with the engines at runup power and brakes locked, activate the boots. Watch the vacuum pressure gauge as the boots inflate. There will be a momentary drop in pressure, but the needle always should stay within the green arc. If it drops below the green arc, either you aren't carrying run-up power on the engines, there's a leak in the system, or there's a hole in the boots. If all is well, visually observe the boots inflating and deflating in the proper sequence. Some systems operate wing, horizontal stabilizer, and vertical stabilizer boots simultaneously while other systems sequence them. Check your POH and assure the system is sequencing correctly. Inflation cycles generally take about 6 seconds, so let them cycle at least three times. This is also a good time to check for *softballing,* a ballooning effect that is caused by the separation of the outer neoprene (the type of rubber used) layer from the undersurface (FIG. 23-4). If caught soon enough, it can be repaired easily.

Boots can also have one of those Gordian knot problems. During preflight everything checks out, but at altitude, some of the boots don't inflate. Indignantly you land, taxi up to the shop, and curse silently as the A&P tells you there's nothing wrong. The problem is subtle, but not really difficult to understand and locate. You almost certainly have pinholes in the boot and they're probably caused by ozone deterioration. These tiny holes are too small to prevent proper inflation on the ground, but inflight, when flying through rain or clouds, the vacuum that holds the boot flat will draw moisture through the holes. Because

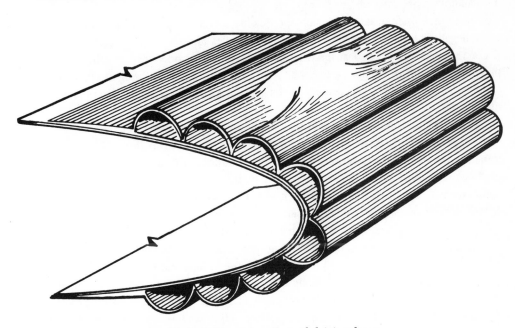

Fig. 23-4. *Ballooning effect of deicing boots.*

few pneumatic systems have water separators, the moisture collects in the valves, lines, and boots. At altitude, the moisture freezes and prevents pressure from reaching the boot, then unfreezes when you descend.

If a boot never inflates, the first thing to look for is a hole. Activate the system and liberally apply a 50 percent soap and water solution to locate the problem. If the bubble routine doesn't work, then you have a more serious problem and it's time to call your A&P. The culprit could be as simple as dirty contacts preventing the flow valve from getting the go-ahead signal. Otherwise it could lie in leaking or broken lines, valves, timers, or relays.

Boots that only partially inflate probably indicate an incorrect pressure-regulator setting. If that's not the problem, then the valves could be stuck in a less than full-open position, the lines could be clogged, or the pump is simply going bad. The easy one to solve is a boot inflating out of sequence—that's timer trouble. On the other hand, boots that inflate when the system is off require some thinking.

When inflight, the airflow over the wing causes a partial pressure around the boots. That is why a vacuum is applied to the boots when they aren't inflated, to prevent the boots from falsely inflating due to the external partial pressure. If there is a leak in the vacuum lines, the ambient partial pressure pulls the boots away from the airfoil in a sort of false inflation. If they automatically inflate on the ground with the system off, you almost certainly have a flow valve stuck in the open position. This is a serious condition, because even partially inflated boots can erode aerodynamic efficiency, making takeoffs and landings more difficult.

Preventive maintenance of pneumatic deicing systems will go a long way toward extending their life. Your A&P can do a lot for holes and tears with a cold patch kit; if done properly, they can last the lifetime of the boot, but don't expect miracles. The process is similar to patching a tire, but you must use a patch kit designed for deicing systems! The possibility of successfully patching a hole is determined by the size and location of the hole.

In the inflatable tube areas, you may apply a patch if it doesn't cross a stitch line (the edges of the inflatable tubes) *and* the hole is ¾ of an inch or less. Holes in non-inflatable areas are equally serious, because the wind tends to tear them open and cause boot failure. Holes up to 3 inches across may be patched, provided they are completely within the non-inflatable area. Remember what I said about miracles, because there are criteria for the maximum number of patches in any given area!

As boots accumulate small surface nicks and scuff damage, you might want to have the surface cover refurbished. This is a sort of "magical" cleaning that actually resurfaces the boot, provided the damage is not deeper than 0.010 inch and there is no air leakage when inflated. It is a protective coating that may be applied more than once, however the manufacturer recommends not doing it more than twice. If the scuffing isn't too bad, it's an economical way of cleaning up the surface and it has the added benefit of greatly improving the appearance.

One word of caution—while most pneumatic deice boots are neoprene, some are estane. If the boots are so labeled, point it out to the A&P before having them refurbished or cold patched, because estane and neoprene are not compatible and a different repair kit must be used.

Let me reiterate a final note on preventive maintenance that applies to propellers, boots, and in fact all rubber parts. The ozone in the air causes premature aging of all natural and synthetic rubber. Sure signs of ozone damage are pinholes, cracks, crazing, and hardening. It is money in the bank to coat all rubber parts with Age-Master No. 1 every 150 flight hours. This is probably the best protection against ozone damage currently available.

For deicing surfaces, follow up that treatment with an application of ICEX. It reduces ice's ability to adhere to the rubber surface, making the deicing system more efficient. The manufacturer recommends applying it every 50 hours to boots and 15 hours to propellers. If you are going to do it after an application of Age-Master No.1, hold off a minimum of 24 hours for curing. You can occupy those hours by using one of the few positive aspects of ice—cooling your favorite refreshments while you contemplate how to stay clean of all the negative aspects of it!

24

Know Your Aircraft's Hydraulic System

Fluids are flexible. They change shape to fit their surroundings, can be divided into parts to work in different places, can move rapidly in one place and slowly in another, and can transmit force in any direction. There are 2 categories of fluid: compressible and incompressible. Compressible fluids (gases) include air and nitrogen; the branch of mechanics that deals with the properties of gases is called *pneumatics*. Incompressible fluids (liquids) include water, oil, and modern hydraulic fluids; the science that deals with the transmission of energy and the effects of flow of liquids is called *hydraulics*.

There are two kinds of hydraulic systems: open and closed. Windmills and waterwheels are examples of open systems. A common example of a closed system is the car rack at an automobile repair shop. The hydraulic systems in aircraft are closed systems because they confine the fluid. In a closed system, fluid pressure can be increased, which increases the amount of work obtainable from a given amount of fluid.

Few pilots can list all of the possible uses of hydraulics in an airplane. One thing that compounds the difficulty of that task is that some potential applications of hydraulics also can be accomplished by the use of electric motors. The more obvious uses of hydraulics are brakes, retractable landing gear, and flaps, but other possible uses include gear struts, engine valve lifters, shock absorbers, nosewheel shimmy dampers, anti-skid systems, and control surface actuators.

THE ROLE OF HYDRAULIC FLUID

Hydraulic fluid is the lifeblood of the system. There are other incompressible fluids, but hydraulic fluid is very special because it allows maximum flow rate with minimum friction and serves as a lubricant for the working parts of the system. Also, it doesn't foam (so it prevents air from entering the lines), it's non-corrosive, and it's compatible with the synthetic seals in the system.

The mineral-based MIL-H-5606 fluid used in virtually all light aircraft can be identified by its red dye. This type of hydraulic fluid is compatible with neoprene seals and hoses; no other fluid should be used in its place. The only significant drawback of MIL-H-5606 is that it is flammable; great caution must be taken while pouring and storing it.

Larger aircraft use a nonflammable, synthetic fluid—MIL-H-8446—commonly known as Skydrol. It is pale purple (some grades are amber or light green) and is formulated for the higher operating pressures and temperatures of larger and faster aircraft. If that sounds like a better product than 5606, don't run out and buy a case—the systems aren't compatible. Skydrol is very susceptible to water contamination, corrodes polyvinyl-chloride (it'll eat the shielding right off of electrical wires), dissolves most aircraft paint finishes and will dissolve the seals used with mineral-based fluid systems. In short, follow the recommendations of your POH.

At best, hydraulic fluid is touchy stuff. Make sure the reservoir cap is fastened securely to prevent leakage and always keep fresh fluid in the system. Old fluid smells sour and the color darkens. Drain old fluid, flush the system with Varsol or Stoddard solvent and replace it with fresh fluid.

Hydraulic systems are very intolerant of contamination; synthetic fluid systems are so bad that the fluid has to be inspected with a microscope. Whenever a hydraulic system component fails, you should drain the fluid, flush the system and fill it with fresh fluid; it's not worth the risk of running contaminated fluid through the system only to have it eat the seals and grind the surfaces.

MECHANICAL ADVANTAGE

The reason hydraulic systems function as they do was first explained by the 17th Century French mathematician, Blaise Pascal. He explained that pressure of static liquid at any one point is the same in every direction and exerts equal force on equal areas. That means that an incompressible fluid can transmit a force—and more importantly, multiply it—anywhere throughout the system (FIG. 24-1). There is a big difference between a static and a moving liquid. A moving liquid is affected by friction, which causes some of the energy to be lost as heat. This problem can be minimized if you know the five major causes of friction: tubing that is too long; excessive or too sharp of a bend in the tubing; too many fittings in the tubing; too high a fluid velocity; and insufficient tube diameter.

SYSTEM COMPONENTS

Most pilots are familiar with the basic hydraulic system that operates aircraft brakes. It is simple, straightforward and easy to understand—but don't let it fool you. Hydraulic

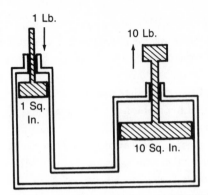

Fig. 24-1. *Because the pressure of static liquid at any one point is the same in every direction and exerts an equal force in all directions, when* F_1 *(1 psi) is applied to a one-square-inch disk, the resultant force,* F_2*, is 10 psi as it acts upon a 10-inch disk.*

systems can be as beguiling and inscrutable as the most complex electrical systems, and while they won't bite you if you touch the wrong place, they have a nasty habit of making a mess, defying common sense, and going awry at the absolute worst possible time.

The heart of the hydraulic system is the pump, which can be operated electrically, by bleed air from a turbine engine, directly off a reciprocating engine, or even by hand. While use of hand-driven pumps in aviation has become relatively rare, some aircraft still use them fairly effectively. The Mooney Mark 21, for instance, uses a single-action hand pump for its flap system; each downstroke pumps the flaps down a little more. There are also double-action pumps, where each stroke up or down moves the flaps. In light aircraft, virtually all major hydraulic systems are engine-driven. The important thing to remember about pumps is that they don't create pressure, they just move fluid. Pressure is created by the fluid's resistance to the movement.

The system reservoir is basically a permanently installed can of hydraulic fluid that contains a reserve supply. Because hydraulic fluid expands as the temperature increases, the can is oversized to accommodate expansion. It also serves to release air from the fluid to prevent the air from entering the lines. Pressurized reservoirs are primarily used on aircraft that operate at high altitudes, because the reduced ambient pressure causes the fluid to foam. Light aircraft use the simpler, non-pressurized reservoir.

Similar to electrical systems, hydraulic systems have fuses and diodes (valves). Fuses are more common in large systems that support several functions. If there is a rupture in a line, the fuse prevents fluid loss and allows continued activation of other hydraulic components. Check valves permit flow in one direction only, preventing fluid from backing up in the system. The pressure relief valve is essentially a pressure-control device. It is set to open at some specific, higher-than-normal system pressure to act as an escape valve. Similarly, the thermal relief valve is activated by excessively high fluid temperature.

Of the many different types of hydraulic systems, the most common is the independent brake system. Older small aircraft use a single-unit diaphragm master cylinder and brake actuator. The master cylinder contains the fluid, and when the pilot pushes on the pedal, the fluid fills the wheel cylinder and applies pressure to the brake. Hydraulic brake systems on newer small aircraft and on all larger aircraft require a reservoir to hold more fluid and to compensate for temperature change. For instance, Piper's PA 28-161 Warrior II has a fixed gear, permitting a very simple hydraulic system. It is highly effective and requires minimal maintenance. The gear operates the toe brakes, hand brake and parking brake. The reservoir, located on the top left front face of the firewall, is accessible for preflight inspection. The system works so well, it is essentially duplicated in other, larger Piper aircraft, such as the twin engine PA 34-220T Seneca III which has a separate hydraulic system for its retractable landing gear.

On the opposite end of the spectrum is the pressurized system (FIG. 24-2). System pressure is maintained by an accumulator which is essentially a metal sphere. The sphere is split in half by a rubber-like diaphragm. On one side is a dry-air precharge, on the other is system fluid. In addition to absorbing system shocks, the compressible air allows pressurization of the system as fluid pushes against the rubber diaphragm. An automatic unloading valve, which senses system pressure, locks the system, trapping the pressure, and reroutes the continuous stream of fluid from the pump back to the reservoir. As system pressure decreases, the valve senses the reduction, opens, and allows the pump to continue the flow of fluid into the system until the pressure builds up again. While effective, such a system is very costly, complex, and has a greater tendency to wear and leak.

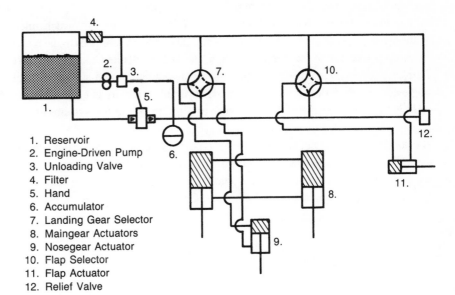

1. Reservoir
2. Engine-Driven Pump
3. Unloading Valve
4. Filter
5. Hand
6. Accumulator
7. Landing Gear Selector
8. Maingear Actuators
9. Nosegear Actuator
10. Flap Selector
11. Flap Actuator
12. Relief Valve

Fig. 24-2. *Pressurized systems are complicated and have a tendency to wear and leak; they rely on an accumulator, which absorbs system shocks, and an automatic unloading valve, which senses system pressure and controls the fluid flow.*

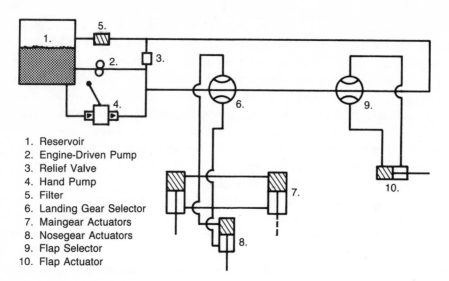

1. Reservoir
2. Engine-Driven Pump
3. Relief Valve
4. Hand Pump
5. Filter
6. Landing Gear Selector
7. Maingear Actuators
8. Nosegear Actuators
9. Flap Selector
10. Flap Actuator

Fig. 24-3. *Open-center systems are the type usually found on general aviation aircraft; they have no accumulator or constant system pressure.*

The open center system (FIG. 24-3) is found most commonly on light, general aviation aircraft. It has no accumulator or constant system pressure. When there is no demand on the system, fluid travels from the pump through the open center of each selector valve and back to the reservoir. When hydraulic power is required, the valve rechannels the fluid to the actuator and the fluid from the opposite side of the actuator goes to the reservoir. This is a much less costly and simpler system and is better suited to light aircraft.

There are several types of basic actuators: single-acting, double-acting, and rotary. The single-acting cylinder (FIG. 24-4) moves under hydraulic pressure in only one direction; the return is a result of some outside force, such as a spring. Some flap systems are single-acting, using the air load to retract the flaps.

The double-acting cylinder (non-differential type) uses the same force in both directions, because each side of the cylinder has identical surface area (FIG. 24-5). Generally, the left rod in the illustration is not connected to anything, but is there to take up space and assure equal surface area (equal pressure) on both sides. Cessna uses the simple rack and rotary-

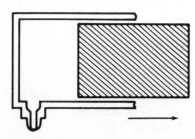

Fig. 24-4. *Single-acting cylinder moves in only one direction; it relies on a spring or other device for return.*

Fig. 24-5. *Double-acting cylinders use the same force in both directions, because the cylinder ends have equal surface areas.*

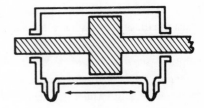

actuator-driven pinion to retract the main landing gear on its single-engine aircraft. For tasks requiring continuous motion, a piston-and-vane type hydraulic motor is used. Such a motor, common on larger aircraft, provides high power output with instantaneous reversal capability.

PREFLIGHT INSPECTION

As with all preflight inspections, follow the POH; however, there are several hydraulic system checks that should be made on all aircraft. Check for traces of hydraulic fluid on the ground, the underside of the airplane, and inside the cowling. Where accessible, check the general security of all hydraulic lines, fittings, and actuator cylinders. Always check hydraulic fluid level and color, especially if it has been changed recently. The wrong fluid will lead to disaster.

PREVENTIVE MAINTENANCE

Remember that hydraulic fluid acts as a lubricant—always check the fluid for proper quantity, type, and quality. It must be uncontaminated. Never change fluid during dusty conditions, reuse fluid, or store extra fluid in an open or dirty container. I cannot overstress the importance of the quality of hydraulic fluid; 70 percent of all hydraulic problems are traceable to the fluid!

Keep lines properly secured. They should not be loose in the clamps. Be especially watchful for chafing of the lines where they pass through the bulkheads or near other components. Fittings should be hand-checked for tightness during preflight.

TROUBLESHOOTING

There are some general principles to keep in mind when troubleshooting any hydraulic system. Because pumps make fluid flow, reduced flow means a pump or pump-drive problem. No flow and no pressure means no pump. Absence of pressure alone doesn't necessarily indicate an inactive pump. Remember, resistance is required to generate pressure. No pressure means no resistance. Find out where the fluid is going, because it's leaking somewhere.

If you have system flow and an actuator doesn't move, only one of three things can be happening: the fluid might be bypassing the actuator through an internal or external leak, the fluid might be returning to the reservoir (means bad seals or a relief valve is stuck open), or there is a mechanical binding somewhere in the system. One of the more common hydraulic system problems is a noisy or chattering pump. There are two likely culprits—*pump cavitation* (sudden formation and collapse of low pressure bubbles resulting from pump rotation) or the pump is drawing air. Cavitation results from too low an operating temperature, a dirty inlet strainer, an obstruction in the inlet tubing, or too high a viscosity fluid. A pump will draw air if there is insufficient fluid, if a leak exists in the intake tube, if there's a bad pump shaft seal, or if the fluid is foaming in the reservoir. There are also a number of actual pump parts that can fail and cause similar symptoms, but low fluid level is the most common cause.

If the system is overheating, first check to see if the heat exchanger is clogged. Otherwise, a relief valve could be operating continuously, creating excessive heat or the wrong viscosity fluid is being used. Slow or erratic pump operation indicates air in the fluid, internal leakage in an actuator, or a bad pump. Low system pressure could be a result of dirty fluid, a relief valve stuck open or the pressure control set too low. Absence of pressure could be the result of a relief valve that is stuck open, a faulty actuator bypassing fluid or, most probably, insufficient fluid in the system.

The last two problems are straightforward, easy to diagnose, and relatively easy to fix. Spongy actuation, most noticeable in brakes, is the direct result of air in the lines. Bleeding the lines of the trapped air will cure the problem. Finally, if the hydraulic pressure gauge needle is bouncing, there is air in the gauge line. By slightly loosening the connection where the line meets the gauge, the air will escape. If all this has sounded devastatingly confusing and complicated, take some consolation from the fact that if you keep the fluid clean and the level up, you can avoid 70 percent of all hydraulic problems.

25

How to Troubleshoot
The Pneumatic System

Aircraft pneumatic systems use airflow to produce either vacuum or pressure for driving gyro instruments, operating deicing boots, maintaining cabin pressurization, and performing other pressure-related chores.

Aircraft of a few years ago were equipped with venturis mounted on the fuselage in line with the propwash (FIG. 25-1). A venturi is basically a tube flared at the ends and constricted in the middle. As the speed of the air passing through the constricted part of the tube increases, the pressure of the air decreases, thus creating a vacuum. The venturi-driven vacuum system proved effective for cruise flight but is highly susceptible to icing.

WET PUMP SYSTEMS

To solve the icing problem, an engine-driven vacuum pump system was developed (FIG. 25-2). The pump, which is impervious to icing, is mounted on the accessory section of the engine and uses engine oil for both lubrication and cooling; hence they are often called ''wet'' pumps. Air from the cabin is pulled through the gyros and into the pump, which is lubricated with engine oil. An oil separator then returns most of the oil to the engine and exhausts the air and any residual oil out of the system.

The system works but it has disadvantages. Because of the need for the oil separator, the system is heavy. It also is complicated; oil sometimes finds its way to the wrong part of the system, contaminating deice boots and control valves. And the oily exhaust air always leaves its mark, streaking the underside of the fuselage.

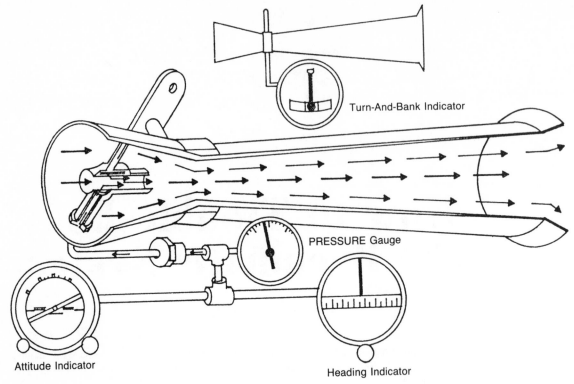

Fig. 25-1. *Venturi vacuum system.*

Probably the biggest drawback of the wet-pump pneumatic system is that instrument flying is limited to lower altitudes. In the low density air of high altitudes, the pump cannot create enough vacuum to drive the gyros. If the wet pump is used in high altitude operations, the gyros and the rubber deicing boots (located on the pressure or downstream side of the pump) could be ruined by oil contamination.

DRY-PUMP SYSTEM

The solution to the many problems of the wet pump came with the invention of—you guessed it—the dry-pump system. The simple, lightweight, dependable and self-lubricating dry-pump system has no oil contamination or cooling problems. Because it can power either a vacuum or pressure system, it is able to drive the gyros and the deice boots as well as pressurize the aircraft door seals.

Vacuum vs Pressure Systems

The most common type of aircraft pneumatic system is the vacuum type (FIG. 25-3). Air from the cabin is drawn into the system through a central air filter. It goes through

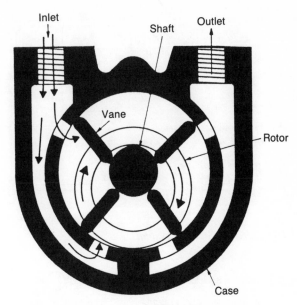

Fig. 25-2. *Vane-type engine-driven pump.*

the gyro instruments, past the vacuum (suction) gauge and through the relief valves. It then goes into the vacuum side of the pump, out the pressure side, through an air/oil separator, and is released overboard.

Twin-engine aircraft vacuum systems work essentially the same, with one pump per engine (FIG. 25-4). This provides a safety margin, because either pump alone creates sufficient vacuum for normal operation. Both pumps connect to a common manifold and share the same tubing and valves.

Vacuum systems are more common than pressure systems in light aircraft because these aircraft generally operate at lower altitudes. But, the vacuum side of the pneumatic pump becomes less and less efficient as ambient air density decreases. Using a vacuum system in aircraft that routinely operate at higher altitudes results in shortened pump life because the pump has to work harder in thin air. Smaller aircraft require less vacuum action from the pump because it usually only powers the gyros.

Pressure systems can move more air at higher altitudes than their vacuum counterparts. Air enters the pressure system through a cabin inlet filter, goes directly into the vacuum port of the pump and out the pressure port to the pressure regulating valve. After going through another filter, the air is routed to the gyro instruments, past the gauge, and is finally released overboard.

Twin-engine aircraft have one pressure pump per engine; as in the vacuum systems, they share common tubing and valves. According to several mechanics, pressure systems tend to create moisture, especially in high-humidity locations—near oceans, lakes, and rivers and in areas with heavy rain. Because moisture in the lines and gyros can cause

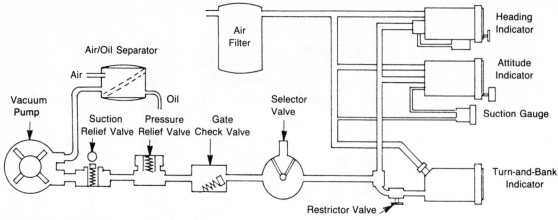

Fig. 25-3. *Typical pump-driven vacuum system.*

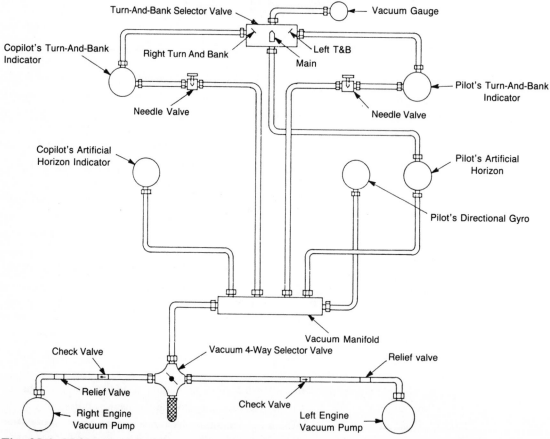

Fig. 25-4. *Multiengine aircraft vacuum system.*

significant problems, filters of aircraft operated in high-humidity areas should be changed more frequently than recommended by the manufacturer.

The dry-air pump is the heart of the modern pneumatic system. The pump rotor has self-lubricating carbon vanes that are specially designed to wear and lubricate the pump at a set rate. Unlike older, bidirectional pumps, the dry pump turns in only one direction. To provide maximum life, the vanes are canted at a carefully calibrated angle, making it essential that the proper pump be installed correctly. A pump that has been dropped should never be put into service. The carbon vanes and rotor are surprisingly fragile, and while there might be no visible sign of damage, internal damage could have occurred.

For gyros to work properly, exact vacuum (or pressure) must be maintained, despite the potential for system fluctuations caused by power surges, deice-boot inflation, and engine-speed changes. The gyro instruments are protected from excessive vacuum by a regulating valve. Under normal operating conditions, the pump draws more air than is necessary to operate the gyros, so the valve regulates the airflow by drawing from an alternate air source that bypasses the gyros. Similar to the vacuum system, the pressure system has a regulation valve that serves to relieve pressure by allowing excess air pressure to escape prior to entering the gyro instruments.

The vacuum gauge needle responds directly to the amount of suction in the tubing that passes by the gauge. In twin-engine aircraft, where there is one pump per engine, the gauge has a pump-failure alerting mechanism. Inside the vacuum gauge—out of the sight of the pilot—two small red balls are held independently by the suction of each engine-driven pump. If, for instance, the left engine fails, its corresponding pump also will fail and there no longer will be suction on the left red ball. A small spring, normally held back by the suction, now pushes the ball into the view of the pilot. This is not necessarily a reliable indication of engine failure. The windmilling propeller of a failed engine could keep the vacuum pump operating normally. In that case, the red indicator would not appear until the pilot feathers the prop!

A pressure gauge works essentially the same as a vacuum gauge in both single- and multiengine aircraft. Both types of gauge should read approximately in the middle of the green arc during normal operations.

Modern fittings and tubing are designed for maximum airflow. While the tubing in older pneumatic systems often had sharp, 90-degree turns, modern systems are engineered with less acute bends. Even the design of tubing itself is geared for minimum flow resistance, because cracks, twists, and crimps in the hose and improper internal dimension can lead to significant airflow problems and seriously decrease the life of the pump.

Pneumatic system filters capture impurities in the air, safeguarding the gyro instruments. According to Don Bigbee, general manager of Aviation Instruments Manufacturing Corporation, clean air is essential for unrestricted airflow, protection of delicate gyro instruments, and pump longevity. AIM, a major supplier of light aircraft gyro instruments, has extensive experience in diagnosing gyro problems, most of which are the result of system contamination. At the head of the list are dust (including women's face powder), cigarette smoke, dirt, some cabin-cleaning agents, and dry fire-extinguishing agents. It is the job of various filters to protect the system from these contaminants.

The inlet air filter is the primary system filter; it removes most major pollutants from the air before they enter the system. The vacuum system has a garter-type foam inlet filter that cleanses the air as it enters the regulating valve. In the pressure system, the inlet filter is located just upstream from the pump itself. An in-line filter, which comes between the pump and the instruments, is used in the pressure system to trap the carbon lubricating particles before they enter the gyro instruments.

PREFLIGHT

Preflight of the pneumatic system is limited to an under-the-cowling check of the general integrity of as much of the system as is visible. Hoses and fittings should appear in good condition, with no kinks or twists. If the pump is visible during preflight, check the base where it attaches to the engine; traces of oil on the accessory drive pad indicate a bad seal. During the cockpit preflight, be especially watchful if the vacuum pump has passed the 500-hour mark; 200 hours for the pressure pump. Be aware of the normal gauge reading for cruise flight; deviations might indicate impending failure. The vacuum/pressure gauge should be an integral part of your instrument scan. Any excessive gyro precession during ground operations is reason to suspect system problems. For instance, the attitude indicator should not show more than a 5-degree bank during taxi turns.

PREVENTIVE MAINTENANCE

The single most effective preventive maintenance a pilot can perform on the pneumatic system is to forbid smoking in the aircraft. Tobacco smoke is a prime cause of clogged filters and shortened gyro life. Where gyros are concerned, small problems rapidly turn into big ones. According to Bigbee, some mechanics check filters at the required time, and if they don't look dirty, they won't clean them. The problem is that dirt gets trapped inside the filter where it's not visible and then works its way out and into the gyros. Literally, any small particle can cause a gyro to fail!

In one general aviation airworthiness alert, the gyros stopped working when the vacuum system had excessive suction. The master air filter appeared clean, but after it was torn open, it was discovered that water had leaked inside. The filter had absorbed the water, swelled and blocked the airflow. From the outside, the filter looked fine.

Bigbee says filters should not be cleaned, they should be replaced. The problem is more subtle than just potential contamination. The longer a filter does its job, the lower the volume of air that can pass through it. If the filter is located upstream from the pump, the decreased airflow forces the pump to work harder. This increase in workload causes the pump to run hotter, which increases the wear on it and shortens its life.

Under normal conditions, the vacuum regulator filter or pressure system inlet filter should be changed every 100 hours of operation or every year, whichever comes first.

Central air filters and in-line filters should be replaced every 500 hours or annually, whichever comes first. If an aircraft is operated in a high-humidity area, mechanics often recommend changing in-line filters every 300 hours.

If you install a new pump, always replace the old gasket with the new one that is provided. Used seals have been compressed and are not designed to be reused. The gasket seals the pump to the accessory-drive-pad oil hole and also serves as a heat shield. A leaking seal will cause the system and pump to be contaminated with oil. A new pump also warrants a check of the system and change of all filters. This should include cleaning the pressure lines from the pump to the instruments, because carbon particles can find their way through the filter and begin to collect there. Also clean the inlet lines.

If there is any drawback to the dry-pump system, it is the potential for carbon contamination. While this problem may be obvious in the pressure system, some mechanics think the vacuum system, with its pump downstream from the gyros, also can be vulnerable to carbon contamination. They hypothesize that as the pump shuts down, the instrument cases momentarily have a lower pressure than the pump, causing a reverse airflow that could pull carbon dust back into the instrument case. Whether or not this actually happens, the best safeguard is to change filters regularly and keep the system clean.

The lining of deteriorated hoses can break away and flow through the system until it clogs a filter, lodges in a valve, or jams the pump. The effect of a collapsed, kinked, or twisted hose is similar to that of a clogged filter. The restricted airflow raises pump operating temperatures, which results in a shorter pump life. Damaged or loose fittings, in addition to acting as another air source that can bypass gyros, can also allow engine-cleaning solvents to enter the pneumatic system. When this happens, the solvent can mix with carbon particles to form sludge. In the pump, sludge will cause the coupling to shear! As a precaution, never direct high-pressure solvents on a pneumatic system component when cleaning an engine. In fact, it is a good operating practice to encase these components in plastic bags before using solvent cleaners. (The bigger the bag, the less likely you will be to miss it on a preflight if you forgot to remove it).

Another problem associated with loose fittings or leaking hoses is the reduced pressure that results in decreased gyro airflow. The pilot, noticing a change in gauge reading, might have a mechanic adjust the regulator. This makes the pump work harder and run hotter to compensate for the air loss, leading to shorter pump life. It is cheaper in the long run to replace bad fittings and hoses than to buy a new pump.

TROUBLESHOOTING

The presence of any of the following trouble signs warrants checking the entire pneumatic system:

1. The pump fails soon after installation.
2. The aircraft has a history of short pump life.
3. The vacuum (or pressure) indication is above or below proper level.

4. Gyro performance is erratic.
5. The deice system malfunctions.
6. A door-seal system malfunctions.
7. An autopilot system malfunctions.

Early pump failure is most often caused by the wrong pump being installed; check the part number of the pump. Otherwise, a thorough investigation of all pneumatic components is in order, including gyros, deice boots and valves, door seal valves, and the pneumatic autopilot system. Contamination can overload a pump, causing it to fail early, sometimes within hours of installation! If there are high, low, or erratic gauge indications, check for hose problems, clogged filters, oil in the system, or loose fittings. If everything else checks out, it could be a simple regulator-setting problem.

Excessive gyro precession is probably caused by dirty filters. If the problem continues after changing them, you almost certainly have a bad gyro. If the gauge indication varies with engine rpm, it is the result of regulator-seat contamination, which prevents constant pressure as the pump speed varies with engine rpm. If you find the mechanic is frequently making small regulator adjustments to correct the gauge reading, the filters are probably becoming clogged. This is particularly common in aircraft that carry smokers and eventually results in premature pump failure.

According to Ralph G. Heysek, administrator of aftermarket sales for Airborne Division of Parker Hannifin Corporation, replacing a pump doesn't necessarily mean the system problem has been cured. Short pump life should be like a red flag to a pilot or mechanic, indicating there is something wrong with the system. He cited the case of a pilot who experienced seven pump failures, one after only 6 hours of operation! Finally, the pilot contacted Airborne personnel directly, and the problem, related to the deice boots, was cured in one day.

Airborne is so concerned with product support that it has developed the Airborne 343 Pneumatic System Test Kit. This suitcase-size unit tests pressure/vacuum levels, regulator and valve settings, and proper operation of pneumatic deicers, inflatable door seals, and pneumatic autopilots—without having to run the engine! Not only is that a tremendous safety advantage, but the system is so quiet you can locate leaking hoses by the sound of rushing air!

Once in a flight, there is nothing the pilot can do about a pneumatic system failure. The simple fact is that pump rotors are designed to break if there is any trouble—otherwise, they might cause an engine problem. While twin-engine aircraft pilots have the edge afforded by redundancy, the single-engine aircraft pilot is backed against a wall in the event of system failure—until now, that is. Precise Flight, Inc., of Bend, Oregon, has developed a standby vacuum system (SVS). This simple system uses engine manifold pressure that is diverted to the instruments by a valve connected to the cockpit with a push/pull cable. The system operates on the differential between the engine manifold pressure and ambient atmospheric pressure. Such a system offers many advantages. With only two moving parts, it is simple and requires no electricity or maintenance. Approved for all Lycoming and Continental engines, its 24-ounce total weight and low price tag make it a highly desirable option. But its best selling point, for the single-engine aircraft pilot, is a peace of mind that never before has been possible.

26

Clear Facts
About
Transparencies

Aircraft windshields, or more properly, *transparencies,* were made exclusively of glass for many years. Even now, glass transparencies have some very significant advantages, such as excellent optical qualities, ease of maintenance, and good resistance to scratching. Difficulty in shaping, greater expense, and added weight are the reasons manufacturers sought alternatives.

"As-cast" acrylic resin transparencies were the answer. Plexiglas, as it is often called, is actually a trademark for one type of single-ply acrylic manufactured by the Rohm and Haas Company of Philadelphia. The Plexiglas formula was patented during World War II, when U.S. fighter aircraft began to sport Plexiglas contour canopies.

As-cast acrylic resin is used in virtually all single-engine and light twin-engine general aviation aircraft. The typical general aviation aircraft comes equipped with a .19- to .25-inch-thick acrylic transparency. The range of thickness isn't very effective in shielding the cabin from outside propeller, engine, and wind noise, so thicker transparencies are available for many aircraft on the aftermarket.

Though as-cast transparencies have no strengthening treatment, they are remarkably strong and resilient. Made of a monolithic polymer, their molecular structure is uniform throughout, so they do not have flaws or discontinuities. And because of their single-ply nature, flaking or delamination of one ply from another is not a problem. As-cast transparencies are inexpensive and hold up fairly well for several years in the outdoors before deterioration and discoloration set in. Acrylic can handle significant temperature changes and is designed to absorb ultraviolet rays, which is especially important in preventing pilot sunburn at higher altitudes.

From a manufacturers point of view, acrylic is molded easily to the contours required by modern aircraft, yet it still maintains a greater transparency than glass—better than 90 percent of visible light through untinted windows. Acrylic has even greater impact resistance than glass. For all of its advantages, acrylic has two significant problems: it lacks hardness, and it is scratched easily (it has approximately the same surface hardness as brass). Many cleaning agents and solvents attack, and some can even eat through, acrylics.

Stretched acrylic, another type of transparency, is stronger than as-cast acrylic and is used on many pressurized and some large unpressurized aircraft. Engineered to withstand greater stresses, stretched acrylic withstands greater temperature extremes, bird strikes as high as 360 knots, pressurization stresses, and inflight hail. The process of fabrication consists of stretching a sheet of thick acrylic into a thinner, specified shape. This stretching makes the acrylic more resistant to cracks and less susceptible to abrasion and crazing. *Crazing* refers to tiny surface cracks that, when the light hits them from certain angles, turn the windshield into a blinding glare.

Another type of transparency is laminated acrylic, in which a layer of vinyl is sandwiched between two layers, or *plies,* of stretched acrylic. The vinyl contains a heating element of fine mesh wire used for windshield deicing. While in theory this is an excellent idea, heating has proved to be the cause of problems. In many units, the result is good windshield deicing, but in some there has been trouble, because acrylic is sensitive to high temperature. Some aircraft transparencies have crazed and even cracked as a result of excessive heat from the deicing system. On some large twins, the captain's side of the windshield is made of glass—which is less sensitive to temperature—and it is equipped with an electric deicing system. The copilot's side is made of less expensive, unheated acrylic transparencies.

THREE KILLERS OF MOLDED-PLASTIC WINDOWS

According to Dick Forler, president of The Glass Doctor (a firm specializing in restoring aircraft transparencies), there are three killers of molded-plastic windows: ultraviolet, water and abrasion.

Ultraviolet (UV) ages and brittles acrylic windows. It turns transparencies a yellowish color that permeates the entire thickness and cannot be removed. UV breaks down the composition of plastic and acrylic transparencies; if left exposed to the sunlight, they have a useful life of approximately two to three years.

According to Forler, a new transparency is designed to bounce a small bird when the aircraft is in flight. This flexibility is evident if you push gently on a new window with your hand; it should flex slightly. An old window, especially one that has spent its life in the sun, could crack easily by doing the same thing. To avoid this problem, store the aircraft indoors.

The use of external window covers, advocated by some, is questionable. Too often a strong wind can cause the covers to beat trapped dust and dirt into the transparency. Those who think they are safe because they use internal heat shields might want to reconsider. Those reflective shields can seriously raise the temperature of the heat-sensitive acrylic. Heat can cause crazing and, in laminated windows, bubbles.

The second enemy of plastic-molded transparencies is water. Forler points out that acrylic has *hygroscopic* qualities (it absorbs water). Transparencies, which actually change thickness all day long, can absorb from 2 to 10 percent of their weight in water, depending on humidity. In flight, as the aircraft builds up a negative static charge, the water acts as a discharge wick. Static electricity literally explodes from the window, leaving a permanent pit. The best preventive maintenance against this water problem is to spend a half hour per week waxing transparencies with paste wax.

The third enemy is your friendly flight-line attendant. Most pilots think these well-intentioned individuals permanently scar windows by using shop rags loaded with metal chips and dirt. The real problem is not damage resulting from foreign objects in the rags, rather it is the liquids they may have absorbed. Solvents, paint stripper, acetone, and especially hydraulic fluid literally destroy acrylic transparencies. One FBO, mindful of this problem, issued a directive that all personnel would not use rags of any type to clean windshields, rather they would use new paper towels of the type found in the restroom. Not only were the towels rough enough to cause damage, they also set up a terrific static charge that strongly attracted dust and other airborne particulate matter, probably causing more damage than the towel's roughness. It is always best to take time to clean the windows yourself. For the same reasons, it is good preventive maintenance to keep the cockpit as clean as possible. A clean cockpit cuts down on airborne dust which, after adhering to the inside of the window, can eventually scar it.

CRACKS, CRAZING AND DELAMINATION

Windows should be inspected regularly during preflight and scrutinized during the annual inspection. In addition to looking for crazing, be especially watchful for cracks; a glass transparency must be replaced immediately if it is cracked. In as-cast acrylic transparencies, cracks often are the result of relief of internal stresses caused by the manufacturing process or installation. These cracks, if caught early, often can be stop-drilled. The pilot should be aware that this procedure does weaken the structure somewhat and the transparency becomes less likely to hold together in the event of a bird strike; potential for inflight structural failure can also be somewhat greater.

Cracks in the outer ply of laminated acrylic transparencies weaken the entire structure, but a strong inner ply is capable of withstanding normal stresses. A crack in the inner ply significantly weakens the structure; such a transparency should be replaced.

Delamination, another problem related to laminated transparencies, occurs when the interlayer separates from the other layers, either glass or acrylic. It is typically caused by deterioration of the weather sealant around the transparency. Water seeping past the sealant gets between the plies. The affected area takes on a cloudy or milky appearance and typically the entire transparency should be replaced. Initially, it is only a problem of visibility; moderate delamination does not particularly affect the bending or tension capability of the windshield. If the visibility is good enough, there is no urgent problem.

Some delamination along the edges of the transparency is normal, as window edges have limited adhesion to allow for temperature expansion. If delamination expands into the window and develops an irregular or jagged boundary, that is an indication of a lack of uniformity of the separation. This situation causes the polyvinyl butyral interlayer to pull chips of glass from the inner glass surface, which will cause failure of the glass ply. If chipping is present or if the problem becomes worse, the transparency must be replaced. Smooth-edged, clear delamination typically does not get worse, as it indicates the original stress causing it has been relieved.

PREVENTIVE MAINTENANCE

The most significant preventive maintenance the pilot can perform on aircraft transparencies is proper cleaning. First, remove excessive dirt with clean, running water. Then clear the transparency with one of the following solutions: mild dish detergent and water; a 50 percent solution of isopropanol and water; pumice and water; or ammonia and water. After applying the solution with a soft, clean cloth, rinse the area thoroughly and then dry it. Use of commercial glass or window cleaners is not recommended. They clean well but contain wax that can cause streaking.

Cleaning acrylic transparencies requires some care. According to Bob Kubichan, manager of product support for PPG Industries, first remove your rings and watch, as they can cause deep scars. Then remove excess dirt from the transparency with a flow of clean water. It is best to locate and remove caked dirt with your fingers while flooding the area with water. Thoroughly clean the crevices around the window framework; dirt hidden there could be dragged out later and scratch the surface. Next, wipe the surface with a clean, soft cloth or sponge using a warm cleaning solution of isopropanol. To remove grease or oil on the transparency, try aliphatic naptha type 2, which can also be used as the general cleaning solution, or either hexane or kerosene. When using the latter two, flush the area with water after the grease or oil is removed. Never use any abrasive materials, strong acids or bases, methanol, methyl ethyl ketone (MEK), or any ammonia-based glass cleaner—all will damage the acrylic. Also be careful about using approved plastic cleaners in aerosol cans. The chemical itself may be both safe and a good cleaning agent, but the propellant chemicals can damage the acrylic.

Choose a cleaning rag carefully. Any of the following are considered safe: 100 percent cotton flannel, 100 percent cotton terrycloth, and genuine chamois (not the synthetic or imitation kind). These rags can be reused if cleaned thoroughly, dried, and stored in plastic bags to prevent contamination from dust. While some individuals advocate the use of paper towels as long as they feel soft when rubbed on your face, others think any paper towel is bad. It is probably best to play it safe and avoid using them. Also avoid using shop towels, even clean ones, and any fabric made of synthetic fiber.

When cleaning the transparency, always rub in one direction, either up and down or side to side. Do not rub in a circular motion as it causes "glare rings." Look at the cloth after you make a pass with it—you have just created homemade sandpaper! Don't rub again with the same section of cloth, because the dirt on it will scratch the transparency. Fold the cloth to expose a clean area of it and wipe again. When finished, rinse thoroughly and dry. Avoid excessive rubbing with a dry cloth as it scratches the surface and builds up a static charge that attracts dust particles. Remove excess water with a chamois, but don't completely dry the transparency; let it air dry.

Once completely dry, polish the transparency with a thin coat of hard wax, such as Vista, Johnson J-Wax, Turtle Wax or Micro-mesh anti-static cream. The latter is designed to remove water spotting and leaves a high-gloss protective finish on the surface. Waxing the transparency prevents pitting by reducing its water-absorption capability, as explained earlier in this chapter.

When the aircraft is being repainted, great care should be taken to protect the transparencies and sealant. Use high-quality masking material to assure no paint product comes into contact with the transparency. An open can of paint stripper should not even be put in proximity of an aircraft transparency. Acrylic absorbs the stripper fumes, resulting in rapid crazing.

In glass transparencies, scratches are often the result of windshield wipers dragging trapped sand and dirt. Clean the blades often and never run the wipers when the window is dry. Light to moderate surface scratches in glass are acceptable until they cause a visibility problem. Light to moderate scratches are usually removable by someone with experience, but done incorrectly the result will be irregular glass removal and distortion.

Scratches in acrylic transparencies usually don't require replacement, unless they interfere with vision. Because windshield wipers are not put on acrylic transparencies, the cause of scratches almost is always physical abuse. There are three types of scratches that can be removed relatively easily. The first is the hairline scratch, usually caused by improper cleaning procedures. The second type is the minor scratch that you can feel with a fingernail. The deep scratch you can feel with the tip of your finger is the third type. Cracks and crazing do not fall into any of these categories.

Crazing, as mentioned earlier, is defined as fine cracks that extend in a network over (or under) the surface of the plastic. Often difficult to discern because they are approximately perpendicular to the surface, they are narrow in width and usually are not more than about 0.001 inch deep. You typically don't see the crazing line itself because it is so small.

RESTORATION

Restoration of crazed transparencies is definitely possible; it can even be done by the pilot, but it is not the typical home project. There are several serious considerations before attempting to resurface your own transparencies. You must know the panel's exact thickness. Ultrasonic measurement is the best method for the novice, but not many novices have access to ultrasound equipment. The existing thickness must be compared to the minimum allowable thickness, which varies from manufacturer to manufacturer; consult the appropriate technical manual. Because the restoration process removes some of the windshield, some transparencies will already be too thin for refinishing. After completion, the transparency must be checked again for thickness and compared to assure compliance with the minimum requirement. Some transparencies may be refinished several times, greatly increasing their useful life.

The second serious consideration before refinishing a transparency is distortion. Varying panel thickness causes disproportion and waviness when the pilot looks through the window. Distortion results from uneven thickness of the transparency from one place to another. To avoid this problem, it is necessary to remove exactly the same amount of material across the entire panel. Another cause is high temperature, above 180 degrees Fahrenheit, caused by excessively hard rubbing—usually from an abrasive-impregnated wheel. The panel thickens as its temperature increases due to friction. Stretched acrylic is made when thick acrylic is stretched thin and formed. This process strengthens the transparency, but if it is reheated, the acrylic returns to its original thickness and becomes weaker again. It is usually more practical for the owner to have transparency restoration done by a professional.

Fig. 26-1. *The MBB BO-105 cockpit had experienced much wear and tear, including being ditched and submerged in saltwater, where it remained for some time. Notice the crazing and deep scratches.*

Micro-Surface Finishing Products, Inc., is a company that does a significant amount of work in acrylic surface restoration. According to its president, Herbert Wilson, the Micro-Mesh process polishes everything from fingernails to metal hydraulic surfaces—everything except glass and ceramics. Micro-Surface will restore transparencies in its own FAA-certified repair shop. They also teach your personnel to do the work at your location. Wilson told me some minor crazing cracks that are small and close to the surface sometimes only need the Micro-Mesh polishing procedure. For more advanced crazing, however, there is a two-part system.

A Micro-Surface technical consultant was called in to help restore an MBB BO-105 cockpit. Because the cockpit was to be used to build a flight simulator that might include a visual system, there was a concern regarding both clarity and distortion of the transparency. Besides the normal wear and tear a helicopter cockpit undergoes, this one had been ditched in saltwater and submerged for some time before being roughly towed back to shore. The assortment of scratches, scars, and marks created a few areas that totally blocked vision (FIG. 26-1). The consultant, using the fairly simple two-part procedure, pointed out that strict adherence to directions was critical. This included a very precise method of rubbing out the transparency. The problem with restoration is not so much one

Fig. 26-2. *A Micro-Surface Technical consultant sands out the damage on the helicopter transparency using a fine-grade sandpaper.*

of conceptual difficulty, rather it is very labor intensive. A Learjet done by Micro-Surface took 22 hours to remove the severe crazing and only 1.5 hours to repolish with Micro-Mesh.

The two-part process includes sanding out the damage and then polishing. The sanding is done with an assortment of both wet and dry sandpapers of progressively finer grades. They are used with an orbital, vibrator-type electrical sander that operates from 12,000 to 15,000 rpm, with an extremely small orbit and no larger than 4×4.5 inches in size. Transparencies with severe crazing or deep scratches, such as the Learjet, take 85 to 90 percent of the total restoration time to remove the scratches. (See FIGS. 26-2 and 26-3.)

The second part of the process uses Micro-Mesh, a cushioned abrasive. Made of individual crystals held in a resilient matrix rather than a hard resin, the crystals give when pushed against the surface. The crystals, either silicon carbide or aluminum oxide, are cloth-backed, so they can be applied by hand or with an electric sander. You cannot push hard enough to cause deep scratching on the surface. Rather, the crystals plane the surface, acting as a sort of uniform cutting tool. This produces a very fine scratch pattern that makes the transparency appear polished. After three days of work, the BO-105 transparency was indistinguishable from new (FIG. 26-4).

Fig. 26-3. *With half of the sanding process complete, it is obvious that the crazing and scratches have been removed.*

Fig. 26-4. *After the Micro-Mesh process is complete, the transparencies of the MBB BO-105 look sparkling and clear—good as new!*

27

See and Be Seen
with
Aircraft Lighting

For quite a few years after the Wright Brother's inaugural flight, night flying was extremely rare. During such flights, pilots were more concerned about colliding with the ground than with other aircraft.

According to Paul Greenlee, aviation lighting consultant and retired technical director for Grimes Division of Midland Ross, a flashlight was typically the only source of light in or on aircraft. As night flying became more common, the flashlight was replaced by a small floodlight mounted on the instrument panel. In the 1920's, night flying became commonplace and once again aviation turned to nautical precedents: red, green, and white running lights were installed on aircraft. The first running lights had 6-candlepower auto bulbs dipped in red or green lacquer. Because these aircraft still didn't have electrical systems, a 6-volt auto battery was installed.

By the 1950's, increased airline traffic required greater aircraft conspicuity. The rotating-beacon anticollision light was installed. Originally a white light, it was quickly covered with a red filter to eliminate light backscatter in the cockpit.

For the pilot, there are two primary Federal Aviation Regulations that apply to aircraft lighting. FAR Part 91.33(c), which describes VFR night flight instrument and equipment requirements, defines required equipment for civil aircraft, including homebuilt and experimental aircraft. It lists approved position lights and approved aviation red or white anticollision lights. Position lights are the red, green, and white running lights that date back to the '20s, but everyone must now conform to the modern requirements of FAR Part 23.1385, ''Position Light System Installation.''

ANTICOLLISION LIGHT SYSTEMS

The anticollision light system is either the traditional rotating beacon or the more modern strobatic lights. Here the aircraft owner must be careful, because the requirements are broken down into four categories, depending on the aircraft's certification date.

Prior to August 11, 1971, all aircraft had red rotating beacons. At that time, the regulation changed again to allow white incandescent lights, which are more luminous than red lights. After July 18, 1977, the regulations were revised again requiring more stringent power and coverage requirements. References in this article are to the latest requirements, however many aircraft types currently flying were originally certificated under older rules. These earlier aircraft, while operating legally, provide minimal "see and be seen" protection in today's heavily populated skies. For safety's sake, all aircraft should have strobes. Simple, cost-effective upgrading is possible for most aircraft by utilizing existing wiring and lighting mounts that are already on the airframe.

While an anticollision light system is almost always a benefit, there can be problems. FAR Part 91.73 states that "the anticollision lights need not be lighted when the pilot in command determines that, because of operating conditions, it would be in the interest of safety to turn the lights off." One instance where lights off would be a prudent decision is IFR flight in cloud or fog. The strobe lights can cause very strong light to scatter around the aircraft, leading to pilot disorientation and nausea. During ground operations, high intensity strobes can temporarily blind pilots of other aircraft near yours, so turning them off is a good idea.

POSITION LIGHTS

Aircraft position lights also are required for night flight under FAR Part 91.73. By definition, that includes any operations from sunset to sunrise. For Alaska, it is flight during the period a prominent unlighted object cannot be seen from 3 statute miles or the sun is more than 6 degrees below the horizon. This same regulation also covers taxiing or parking aircraft and requires that either the area be illuminated well, the area is marked by obstruction lights, or the aircraft have lighted position lights.

According to FAR Part 23.1385 (deals with position light system installation), the forward red and green lights should be spaced as far apart as practicable, typically on each wingtip. The lights face forward with the red light on the left side and the green light on the right from the pilot's point of reference. The rear-position white light should be mounted as far aft as practicable; usually on the tail, though some aircraft have aft facing wingtip lights instead.

Because the relative position of the lights helps pilots of other aircraft determine your direction of travel, the lights' field of coverage is carefully specified. Vertical pattern must be 180 degrees centered on the horizontal centerline (FIG. 27-1). The horizontal pattern for the red and green lights is 110 degrees of coverage each from the centerline, and it's 140 degrees for the white light. With this standardization, pilots of other aircraft can

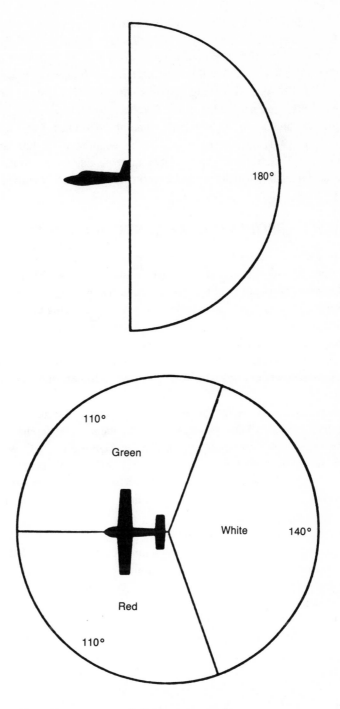

Fig. 27-1. *Position light pattern requirements.*

determine your direction of flight. The phrase used by many pilots to help remember the proper orientation is "red, right, return" which means if you see an airplane with a red light showing to your right side, then it is traveling toward (returning to) you.

Specifications for anticollision light systems (rotating beacon or strobes) are contained in FAR Part 23.1401 Anticollision Light System (FIG. 27-2). The system, which can include one or more lights, must project lights 360 degrees around the aircraft's vertical axis and 75 degrees above and below the horizontal plane of the aircraft. For aircraft certificated after August 11, 1971, the system must produce a minimum of 400 candlepower in a forward direction. The effective light-intensity requirement diminishes as the angle increases from the centerline; tables within the regulation provide the tolerances. The flash rate should be between 40 and 100 cycles per minute.

PROBLEMS ASSOCIATED WITH STROBE LIGHTS

Advisory Circular (AC) 43.13-2A, "Aircraft Alterations," expresses concern about direct or reflected anticollision light rays interfering with crew vision. While it cautions pilots to be sure that proper field of coverage requirements are to be met in accordance with FAR Part 23, it suggests acceptable methods of preventing crew vision problems. The favored method is to paint a non-reflective black paint on the back of propellers, nacelles, and wing surfaces. It then suggests checking its effectiveness on the ground during a night preflight. This is the preferred method, as in flight is a bad place to discover problems. Pilots can experience spatial disorientation, headaches, and nausea as a result of interference.

Communication and navigation interference should also be avoided. Capacitor discharge lights (strobes) can produce radio frequency interference (RFI). This is particularly true in low-frequency equipment such as ADF and Loran C, where it manifests itself as audible clicks in the speaker or headphones. RFI doesn't actually disrupt signals, but it can be very annoying and distracting to listen to. Good systems already are shielded and engineered to minimize, if not totally eliminate, this phenomenon. In most cases it is a matter of improper installation.

Some simple installation guidelines should keep your radios quiet. Most important, always locate the strobe power supply a minimum of 3 feet from any antenna, especially one for a low frequency radio. Be sure the lamp-unit (flash tube) wires are separated from any navigation and communication radio wiring, even input power. Running wires together in a bundle through the wing is a sure way for RFI to set up disturbance in wires that are close to each other. Be sure the power supply case is adequately bonded to the airframe. In fact, for the wires that connect the lamp to the power supply, their shielding should be grounded, but only at the power supply end. Incidentally, this phenomenon should not be confused with the normal audible "wheeeep . . . pop" tone of the strobes' power supply. You should be able to hear this sound with the unaided ear when the engine is shut down.

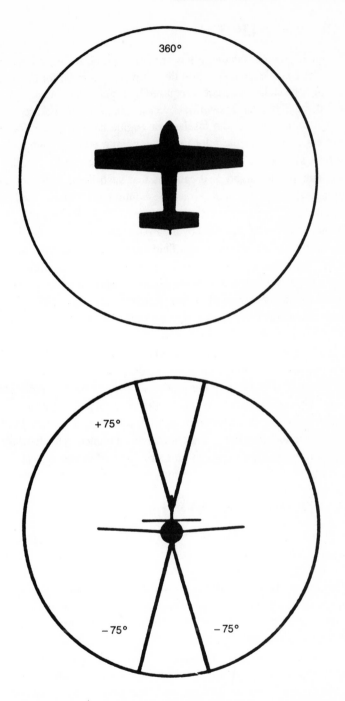

Fig. 27-2. *FAR 23.1401 Strobe light projection requirements.*

OTHER AIRCRAFT LIGHTS

Despite popular opinion, landing lights are not required for most aircraft. FAR part 91.33 only requires a single landing light if the aircraft is operated for hire at night, either VFR or IFR. Only one additional light is required by regulations—an icing light. Aircraft that are certified for flight into known or forecast icing are required to have one. This incandescent light is installed in the fuselage or engine nacelle and is focused over the wing's leading edge to permit the pilot to monitor wing icing at night.

AC 43.13-2A also refers to "supplementary lights that can be installed in addition to position and anticollision lights." It cautions that such lights must not interfere with required lighting. Anticollision lights must be continuously visible and unmistakably recognizable and their conspicuity may not be degraded by supplementary lights. While such lights are not specifically regulated by the FARs, they must be approved by the FAA as a supplemental type certificate or have a field approval by an FAA inspector. Though the types of supplemental lighting systems proposed have been limited only by the imagination of designers, only a few have gained widespread acceptance.

Special recognition lights were developed primarily for aircraft with landing and taxi lights located on retractable landing gear. Recognition lights are high intensity white lights mounted on the wingtips with a forward, and often sideways, projection.

Vertical stabilizer surface illumination lights, also called "logo lights," are particularly popular with the airlines. According to Greenlee, the practice of illuminating the vertical stabilizer was started by TWA during the 1950s. As airports became more congested and traffic increased, it became very easy to lose aircraft position lights among the proliferation of colored lights. Stabilizer illumination helps both the tower and other pilots recognize your airplane; the aircraft tail is lit by small floodlights mounted in the fuselage, stabilizer, or nacelles. According to DeVore Aviation Corp., manufacturer of Tel-Tail Lights, it also helps pilots in flight discern your aircraft's direction of flight, altitude, and speed.

VISUAL CONTACT LIGHT SYSTEM

Anyone who has ever looked for traffic at night over a city will appreciate a product offered by Precise Flight, Inc., of Bend, Oregon. The Visual Contact Light System (VCLS) or Pulselite, makes the aircraft stand out even with a background of red tower lights, city emergency vehicles, and other aircraft strobes and beacons. Landing, taxi, and recognition lights have a stationary appearance when viewed head-on, at all airspeeds. Identification and recognition of head-on aircraft can be difficult until they are so close they already have become a serious hazard.

The VCLS pulses any of those steady rate lights already existing on an aircraft at a rate of 40 to 90 cycles per minute. It is most effective when it alternately pulses lights, for instance a landing light on the left wing then one on the right wing, or a nosewheel landing light, then the two wingtip recognition lights. The combination is dictated by the placement of lights on your aircraft, but the effect in any case is that the lights appear to be moving when viewed head on. It's definitely an attention getter!

According to Precise Flight, it also increases lamp life by as much as 20 times by eliminating the major limiter of bulb life: *shock cooling*. The Pulselite, which flashes the lights between 30 and 90 percent of their normal intensity, never allows the lamp to get "dead cold," meaning the bulbs don't actually extinguish, they just fade. An additional advantage to this system is that the current load is equivalent to that of only one steady lamp. The system is timed to unload one lamp as the other loads, so there is minimal demand on the electrical system. From an operational point of view, Pulselite is fully automatic and quickly switched from pulse to steady rate to off, leaving the pilot's hands free for more important matters. Bird strike research has turned up an extra bonus: reduce bird hazard. It seems many types of birds avoid flashing or pulsing light!

PREVENTIVE MAINTENANCE

Preventive maintenance of lighting systems requires some attention to detail. The plastic enclosures that cover position lights can alter the light's output from 5 to 30 percent, depending on the clarity of the molded plastic and the radius of the curve through which the light is projected. Another area to be cautious about is replacement of position light bulbs. There are three approved forward position light bulbs available for aircraft: 21, 26, and 37 watts. The 21-watt bulbs won't meet minimum intensity requirements at practical line voltages. The 26-watt bulb requires a minimum of 13.2 volts at the light assembly to comply with the intensity requirements, but this does not allow for plastic enclosures such as an enclosed wingtip. With as much as a 30 percent reduction possible, it would be easy to drop below minimum intensity requirements. The 37-watt bulb will comply with 12.5 volts and provide a much brighter light than required at normal line voltage.

According to Whelen, the best preventive maintenance for strobe light systems is to use them. Power supply longevity is a result of regular use as it improves their proper functioning. When left unused for a long time, the electrolytic condenser loses polarity formation. Power supplies out of service for a year or more have a great potential for system failure. Such power supplies should be removed and operated at a voltage reduced by 25 percent for 10 to 15 minutes before putting them into normal service. This prevents overheating of the condenser while it reforms after its long period of inactivity.

Caution must be exercised when dealing with strobe light systems, because they are high voltage devices. For one thing, any reversed polarity of input power causes permanent damage to the power supply. While the problem might not always be immediately apparent, it will cause premature failure. Also, the condenser builds up and stores approximately 450 volts dc, a very nasty shock for any unsuspecting person who grabs it.

The other end of a strobe light system—the xenon flash tube—has some special consideration too. A good, quick check of the system can be had by listening to the flash tube using a paper cup as a stethoscope. If the power supply and trigger transformer are good, you will hear the trigger spark snapping like a spark plug. In fact, you can feel it by putting a finger near the strobe. Don't worry—it's a very low energy pulse and won't have any harmful effect.

TROUBLESHOOTING

There are a few conditions that require replacement of a tube. Xenon tubes are highly photosensitive, so if you find a strobe will fire normally when exposed to external light but not in darkness, it's time for a tube change. Similarly, the flash tubes grow harder to fire as they get older, especially with exposure to very high temperatures. In that situation, they might fire when the engine is running, but not on the lower battery power; again, it's time for replacement. The tube itself is airtight and occasionally a leak will develop as a result of eggshelling of the glass or partial failure of the seals where the wire enters the glass. This condition is inevitable as it is caused by the hot-and-cold cycling of the system. Everything gets old and wears out.

The other cause for replacement is *self-ionization*. When one or more strobes begin to glow a continuous light blue, the entire system will become inoperative. This typically occurs when the system voltage is highest. To verify the problem is ionization, turn off the system, wait a few minutes, then turn it back on. If everything operates normally for a minute or so, then fails and you see the glow, you've got self-ionization. Replace the affected xenon tubes. When replacing an old xenon tube, you might discover other tubes begin to misfire or skip. In most cases, this signifies they too are getting near the end of their service life. To check them out, remove the tube and operate it at 20 percent of its normal input power supply. If it operates at the reduced level, it still has good service life left. Another problem, found only in double-flash systems, is an intermittent second flash. Such systems are designed to operate at normal line voltage. What is probably happening is the system voltage is dropping below the battery charging voltage, and the second flash isn't getting off. Other electrical systems problems are probably to blame rather than the strobe system.

PART 4

Instrumentation

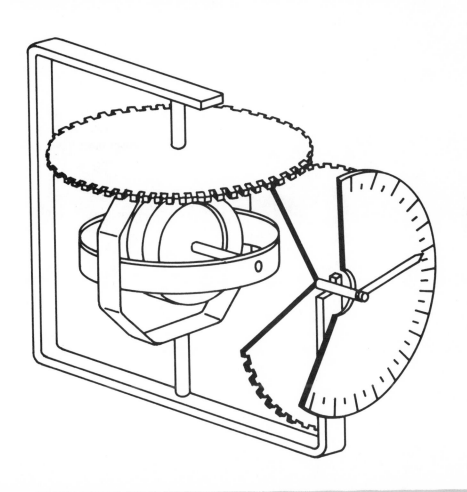

28

How to Troubleshoot the Gyro Instruments

Of the three gyro instruments (attitude, heading, and turn), the most heavily relied upon is the attitude indicator (FIG. 28-1). With the gyro remaining fixed relative to the earth, the instrument displays true vertical and horizontal information regardless of the attitude of the aircraft or its turn rate. The horizon line is fixed to the gyro, so it always parallels the earth's surface. Older attitude indicators require setting (caging) the gyro after engine start. To do this, the pilot pulls the caging knob, which forces the two gimbals into vertical and horizontal positions, orienting the gyro with the horizon. Upon releasing the knob, the gimbals are free to rotate, allowing the gyro to remain parallel to the earth's surface.

There are several problems associated with the attitude indicator, primarily in the form of bank-and-pitch errors. They are most significant during shallow banks and after 180-degree turns. The instrument will indicate slightly less than the actual bank of the aircraft, and after rollout from a 180-degree turn, it will indicate a slight bank in the opposite direction. The instrument will also be slightly off on pitch, causing the uninformed pilot to fly the airplane into a shallow descent. It is imperative that the attitude indicator is always substantiated with a good cross-check of other instruments. Fortunately, the errors are very small, and a built-in erecting mechanism can correct them quickly once straight-and-level flight has been reestablished.

There are also acceleration and deceleration errors. As the aircraft accelerates, the attitude indicator tends to give an erroneous, slightly nose-up indication; deceleration has the opposite effect. It is important to realize that the attitude indicator only *approximates* actual aircraft attitude, however pilots have traditionally relied excessively on the

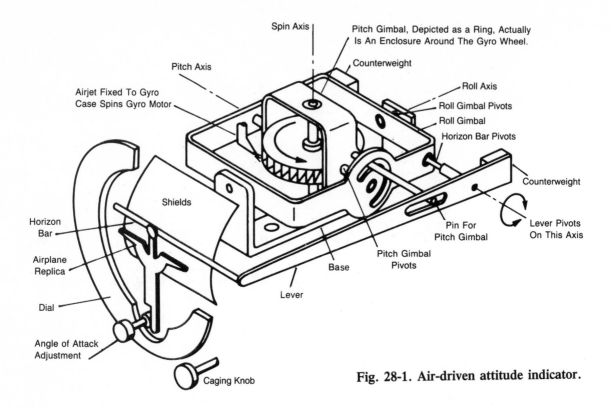

Fig. 28-1. Air-driven attitude indicator.

instrument, occasionally finding themselves helpless without it. Several years ago while flying over Atlanta on a "severe clear" morning, I chanced to overhear a student pilot declare an inflight emergency. When asked, the student somewhat shakily explained his attitude indicator had just failed and he was lost over the city on his first solo flight. The controller in the tower had to explain to him that he could parallel the airplane's wing struts with the ground to make turns and he would be safe.

The heading indicator, which is not a magnetic (north-seeking) instrument, must be set to a compass (FIG. 28-2). The gyro then holds its position while the aircraft literally moves around it. The instrument does need to be reset periodically, due to both random and apparent drift. Random drift is caused by bearing friction and slight imbalances in the gyro and its gimbals.

Apparent drift is caused by several things. The rotation of the Earth is responsible for some of the apparent drift. At the equator, there is zero effect but as the aircraft operates farther and farther away from the equator, the drift increases until at the north and south poles there is as much as 15 degrees of drift per hour.

A second cause of apparent drift is the aircraft changing positions over the Earth. To minimize this, the instrument technician balances the gimbal rings to compensate for

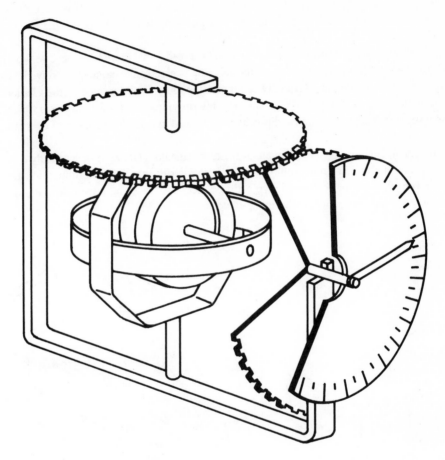

Fig. 28-2. *Heading indicator.*

local drift. If the aircraft is flown to the opposite hemisphere or even a different part of the country, the error could become quite pronounced. Changes in excess of 50 degrees latitude necessitate in recalibration of the instrument.

It is worth noting that checking for precession (drift) is not quite as simple as one might think. The average pilot will set the heading indicator before takeoff and check it against the magnetic compass 15 or 20 minutes later. Based on the comparison, a judgement as to the instrument's accuracy is made, which is like comparing apples and oranges, as the saying goes.

To accurately check for gyroscopic precession, the pilot should turn the aircraft to the same heading used to set the heading indicator originally. This is because compass deviation varies with heading; the error you are seeing could be the result of a different compass deviation at the present heading.

There is also compass variation, the result of crossing isogonic lines. If you are flying cross-country, this should definitely be taken into account when calculating the amount of precession.

For the pilot fortunate enough to have a synchronized gyro, there is no need to worry about precession. The gyroscopic heading indicator is electromechanically "slaved" to a magnetic sensing element. Remotely mounted (typically in the wing tip, the element is isolated from local magnetic disturbances. This provides a constant magnetic update to the gyro, preventing precession and precluding the need to reset the indicator after engine start.

The grand-daddy of gyro instruments is the turn indicator (FIG. 28-3). It is usually driven by a power source different from the other gyro instruments. If the attitude indicator is air-driven, the turn indicator will typically be electric. On some newer aircraft, one might be powered by ac, while the other is dc.

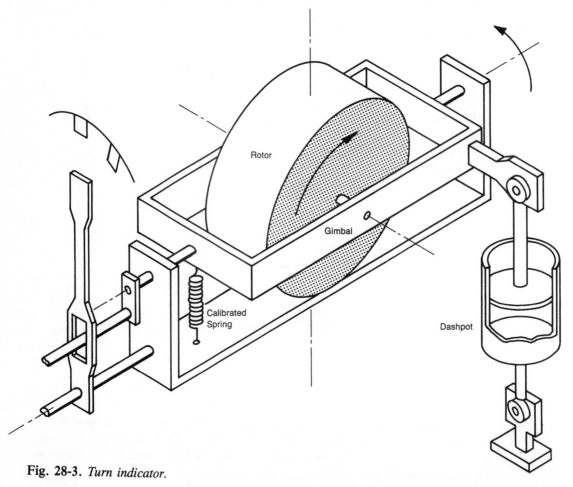

Rotor

Gimbal

Calibrated
Spring

Dashpot

Fig. 28-3. *Turn indicator.*

The most common turn indicator is the turn and slip, or what many call the "needle/ball." This instrument actually senses rotation about the vertical axis (yaw) only. The turn must actually be in process before it is sensed! The instrument is calibrated to a standard rate of turn. In light aircraft, this rate is 3 degrees per second, producing a complete circle in 2 minutes. Faster aircraft have instruments calibrated at 1.5 degrees per second for a 4-minute turn.

The slip indicator is actually an inclinometer and is not connected to the gyro system at all. It merely indicates if the amount of rudder is correct for the angle of bank (coordinated). The ball, usually made of black agate or steel, is placed in a curved, kerosene-filled glass tube. Since the ball is heavier than the liquid, it rolls to the low point of the tube. The liquid provides a dampening action to slightly resist the ball's rolling tendency. In straight flight with the airplane leaning to the right, the ball rolls to the right indicating a *slip* (the result of one wing being low or one engine having less power). In a coordinated turn, centrifugal force and gravity combine to produce an *apparent* gravity, so the ball stays in the center. If there is excessive bank for the right turn, the ball rolls to the right (low) side; if there is insufficient bank, the ball rolls to the left (outside). To assure a coordinated turn, the pilot should keep the ball in the center at all times. The saying is "step on the ball," meaning if the ball is displaced to the right, apply more right rudder pressure, and if it is to the left, use more left rudder. It is also possible to decrease the bank and achieve the same results, for instance in a situation where the aircraft was already banked excessively.

A newer type of turn indicator that is replacing the turn and slip is the turn coordinator. This instrument senses rotation about the vertical (yaw) axis and the longitudinal (roll) axis. As soon as one wing becomes lower than the other, it senses the turn and immediately displays it to the pilot. This makes it easier for the pilot to coordinate with rudder.

RIGIDITY IN SPACE

All gyros operate under the same physical properties: *rigidity in space* and *precession*. The faster a gyro spins, the greater is its tendency to keep its spin axis in the same direction. Quite literally, the gyro inside the instrument maintains a constant position while the airframe moves around it. It is this "rigidity in space" that provides a fixed point of reference for the pilot whose senses are easily mislead when outside visual contact is lost.

The gyro wheel has its bearings in a ring called a *gimbal*. Gyro instruments are gimbaled to permit the gyros to move in specific directions and prevent motion in others, depending on the purpose of the instrument. The force that causes the gyro to spin can be either air or electricity. Older aircraft, before electrical systems, used a venturi mounted outside the airplane as a source of vacuum. While it was Bernoulli's principle in its simplest form, there were some significant problems. It was designed to operate at cruise, approximately 100 mph, which prevented any form of preflight check or calibration until after the aircraft was airborne!

The venturi, which operates fundamentally the same as a carburetor, suffered from another formidable problem: its high susceptibility to inflight icing similar to carb ice. There was also the problem of potential foreign object damage. As aircraft became more sophisticated, the venturi was replaced with an engine-driven vacuum pump which is still in use today. Some aircraft use a pressure system rather than vacuum, but they essentially operate the same way.

The vacuum pump draws air from the gyro case, causing a partial pressure. The opposite side of the gyro case has an air inlet that allows cabin air, passing through one or more filters, to enter the case through a small jet nozzle. The jet directs the airflow toward the gyro's rim where there are buckets (vanes) that catch it, causing the gyro to spin much like a waterwheel (FIG. 28-4). The optimum speed varies from 8,000 to 18,000 rpm depending on type of gyro and manufacturer and is controlled by the vacuum pressure setting. Because the reliability of the instrument depends on the gyro spinning at the prescribed speed, it is important that the vacuum be set correctly. There is a controlling device in the turn indicator to reduce the vacuum, so the pilot should expect to see gauge readings appropriate to the attitude and heading indicators. In the electric-driven gyro, the rotating mass of the gyro is typically the rotor of the motor itself, forming a neat, compact package (FIG. 28-5). Ac-driven systems operate at speeds as high as 24,000 rpm.

PRECESSION

A rotating gyro resists any force trying to alter its spin axis; that rigidity in space is what makes it useful to flying. The spin axis will move *somewhat*, but because of the rotation it moves 90 degrees to the applied force in the direction of rotation (FIG. 28-6). In the gyroscopic heading indicator, this anti-productive force causes the heading to drift slightly. Even in straight-and-level flight, there will always be some precession due to bearing friction, but excessive precession indicates an incorrectly operating instrument and is unreliable.

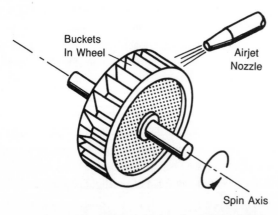

Fig. 28-4. *Jet nozzle and buckets.*

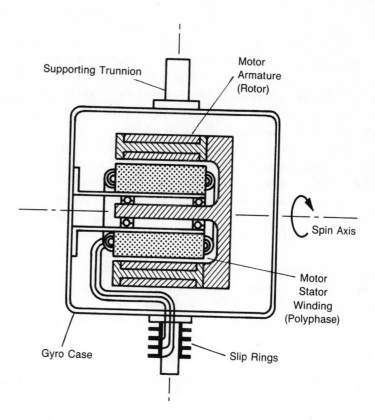

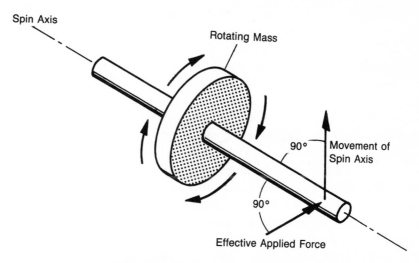

Fig. 28-6. *Gyroscopic precession.*

PREFLIGHT

It is important to allow sufficient time for gyro instruments to spin up to the correct rpm before relying on their accuracy. For air-driven instruments, that means at least 5 minutes; electric-driven, 3 minutes. This is especially important during cold weather operations when lubricant and contamination within the instrument can form a sludge that can significantly resist spinning.

The attitude indicator should be monitored during taxi. It should not display more than a 5-degree change in pitch or bank unless you are taxiing on very hilly terrain. There was a time when instructors taught students to make an abrupt stop while taxiing to cause the airplane to sharply pitch down. During this maneuver, the pilot was supposed to watch the attitude indicator and see if the pitch changed accordingly. Such a procedure should be avoided, as any hard braking is bad for the aircraft and can cause significant bearing damage to the gyros.

The miniature airplane on the attitude indicator is adjustable vertically, allowing the pilot to calibrate it as a reference for straight-and-level flight. Leave it alone when you are not flying. The previous position will probably serve as a good guide for the next flight, at least until after level-off when you may then recalibrate it. Because airspeed, load, and ambient conditions vary the exact pitch setting, it is essentially impossible to set it correctly on the ground. The procedure is to establish straight-and-level flight by reference to airspeed and altitude, and then calibrate the miniature airplane. Changes in airspeed require recalibration. Many pilots set it for straight-and-level cruise and use it as a reference for all other configurations. For instance, one bar-width up can produce a 500-fpm climb at cruise power and one bar-width below a 500-fpm descent.

Prior to an instrument takeoff, never do a fast 90-degree taxi turn. Both the heading and attitude indicators will precess, providing unreliable information during one of the most critical phases of flight. This precession lasts about 1 minute and can be avoided by taxiing slowly or waiting a minute while lined up with the runway before takeoff.

Preflight of the heading indicator includes setting it to a reliable magnetic reference. This should be done while on the ground, preferably prior to taxi. Once set, the pilot should observe it during taxi to assure it is changing heading appropriately. Don't taxi, or fly, with the heading indicator caged—it can damage the bearings.

Use the caging mechanism only to reset the gyro in straight- and-level flight or when stopped, on the ground. If the heading immediately changes after setting, the instrument is unreliable. However, remember the instrument needs approximately 5 minutes after start-up before setting the heading to assure it is up to speed.

Many pilots use the runway heading to calibrate the indicator, but the runway number is not an accurate representation of actual runway heading, varying as much as 5 degrees. Instead, a properly calibrated, undisturbed magnetic compass should be used. Prior to taking the reading, make sure there are no pens, watches, stopwatches, or other magnetic articles laying near the compass that could introduce errors.

If you happen to have a gyro system that operates from a venturi, it is not accurate enough to check or set the instruments while on the ground. Venturi systems are essentially unreliable until you are in flight. This presents an additional problem to the instrument pilot, because to accurately set the heading indicator, the magnetic compass must be settled in straight-and-level flight—uncommon conditions immediately after takeoff into bad weather!

All gyro heading indicators not slaved to a magnetic compass must periodically be compensated for drift. Drift error in excess of 3 degrees per 5 minutes signifies impending gyro failure, and the instrument should be considered unreliable. The gyro should be reset in unaccelerated, straight-and-level flight while the compass is settled down.

You should allow the turn indicator 5 minutes to spin up to the correct speed before relying on it (3 minutes for electric systems). Observe it during taxi turns. The needle should indicate correctly while the ball moves to the outside of the turn.

PREVENTIVE MAINTENANCE

The single most important thing the aircraft owner can do to extend gyro life is to keep the air filter clean, according to Don Bigbee, General Manager of Aviation Instrument Manufacturing Corporation (AIM). He takes issue with the idea of cleaning the filter every 4-500 hours, "don't clean it, change it!" he told me. Then he hastened to add that if you operate in environments where there is a lot of dust, smoke, or smog, change them even more often. According to Bigbee, the filter might look clean, but dirt gets trapped inside, works its way out, and goes up the lines to the instruments. He suggested that when changing the filter, you also check the lines going out of the filter to see if they are dirty. If so, then the whole line is dirty and needs to be cleaned. The problem is if any small particle finds its way into the gyro, it can lead to failure. Most pollution is human-induced, with cigarette smoking probably the biggest culprit. Other problem causers are: woman's face powder and cabin cleaning materials. The simple fact is that any airborne particulate matter can eat up gyro bearings.

Gyros are also extremely vulnerable to shock damage. One instrument repair specialist told me the part most often damaged is the bearings, and the main cause is shock. Gyros should only be handled by qualified personnel, and they should always be transported in shock-absorbent containers—even from the airplane to the shop!

AIM has an enlightening handout on destructive shock to gyros. It discusses the amount of Gs involved in dropping one. For handling purposes, a G is defined as the dropping distance divided by the stopping distance. For instance, if you drop a gyro from a height of 1 inch onto a glass table (the glass is assumed to flex .005 inches), the formula would be $1/.005 = 200$ Gs! A fall of only 1 inch would certainly cause the gyro bearing races to deform when the ball bearings press against them, leading to an early failure. If the gimbal bearing takes the shock, it will lead to excessive, sporadic precession.

New gyro instruments should be kept in their original shipping containers as long as possible. Use the same container for transportation to other locations or when returning it to the manufacturer for repairs. Always use the protective wrap provided, and do not remove the indicator cover anywhere other than a designated clean area of a certified repair facility to prevent instrument contamination.

When removing a gyro instrument from the airplane, wait at least 10 minutes after shutdown to assure the gyro has stopped spinning—never remove a gyro while it is still running. Gyros are so sensitive to shock that other objects should not be allowed to bump against the gyro either. A good rule of thumb is, "if you set a gyro down on a hard object and you can hear it, you have probably set the gyro down too hard." Reputable repair shops always work on gyros with a cushioned pad underneath.

In the airplane, gyro instruments should be "caged" prior to doing aerobatics, especially attitude indicators, if the instrument has a caging mechanism. The pilot should always avoid abrupt braking of the aircraft. Deceleration places heavy loads on gyro bearings—another good reason for a carefully planned approach and minimum braking on the landing roll.

TROUBLESHOOTING

Heading indicators with excessive drift (more than 3 degrees in 15 minutes), turn indicators with sluggish response, and attitude indicators that are slow to erect and/or show inappropriate deviation from level flight could be suffering from any one of several problems. Probably the most common cause is worn bearings; it's the friction that causes excessive precession.

Such symptoms might also indicate insufficient power to the gyro. In an air-driven system, this could be an inoperative pump, improperly set regulator, leaking pneumatic lines or fittings, or an obstruction in the venturi. With an electric gyro, the culprit is probably low system voltage caused by an inoperative or malfunctioning generator. Check the vacuum gauge (if air-driven) or the ammeter/warning light (if electric). If those appear normal, inspect the appropriate connections to the back of the instrument case.

If you have the old venturi system, the first thing you should do is check the venturi for blockage. Worn or damaged bearings make themselves known in another way—gyro whine. If you listen carefully right after engine shutdown, you will hear the still rapidly spinning gyros whine if the bearings are going bad.

Excessively low suction gauge readings could be the result of the pressure regulator setting being too low or a leak in a line or fitting. High readings could be the result of a high pressure-regulator setting, but be careful, it could be a clogged filter. Think of a straw in an extra thick milk shake. If the end of the straw clogs up, the harder you draw, and the lower the pressure is in the straw until it collapses. So in this case, the worst thing you could do would be to lower the pressure regulator setting—even less air would get to the gyro. A good rule of thumb is never lower the pressure regulator valve setting without first checking the filter.

Table 28-1. Vacuum Gauge Readings

Instrument	Minimum	Desired	Maximum
Attitude indicator	3.5	4.0	5.0
Heading indicator	3.5	4.0	5.0
Turn indicator	1.8	1.9	3.2

SAFETY PRECAUTIONS

The single most important safety precaution is to know the system in your aircraft. Know where appropriate circuit breakers are, have a plan for saving battery power to operate the most essential instruments, and know which instruments are driven by electricity and which by vacuum or pressure. For instance, if a vacuum pump shaft failed during an instrument takeoff, the air-driven gyros would slowly wind down. More than one pilot has followed an inoperative attitude indicator *down*. Always use a cross check to assure proper instrument operation including VSI, altimeter, airspeed, and a gyro instrument of a different power source. Believe what the majority of the instruments are telling you. If one instrument disagrees, it is probably wrong. Every few scans should include a check of the vacuum (or pressure) gauge and ammeter. See TABLE 28-1.

Prior to entering actual IFR conditions, observe the gyro instruments carefully to assure correct operation. Monitor the vacuum gauge and ammeter. The time to discover problems is before you are forced to compensate for them.

Tumble limits on older heading indicators are 55 to 60 degrees of roll and pitch, but the newer horizontal card indicators are good from 80 to 85 degrees. Heading indicators do have gimbaling error on some headings, during banked turns. Typically, there is a 2-degree error at 20 degrees of bank, a 4-degree error at 30 degrees of bank, and a 10-degree error at 45 degrees of bank. Gimbal error disappears upon return to level flight.

Older attitude indicators have tumble limits between 100 and 110 degrees of bank and 60 to 70 degrees of pitch, while the newer electric 3-inch horizons typically have 360 degrees of roll and 85 degrees of pitch up or down.

A case of tumbling gyros is bad news; it can cause significant damage to the gyro and in instrument conditions can be disastrous for the pilot. Be prepared for possible gyro failure or accidental tumbling. Your best insurance policy is to maintain partial panel proficiency and remember the old faithful turn-and-slip indicator is your most reliable gyro because it won't drift or tumble!

29

Know Your Plane's Pitot-Static System

Of the five primary flight instruments (airspeed, altimeter, vertical velocity, compass, and turn indicator), the first three use the pitot-static system. These instruments interpret aircraft performance within an air mass.

AIRSPEED INDICATOR

Regardless of the level of sophistication, all airspeed indicators share one thing in common: they compute dynamic pressure by measuring the difference between air pressure resulting from airplane movement through the air (ram) and ambient air pressure (static). A good example of ram air can be felt when you put your hand out the window of a car traveling at 50 mph. Static air pressure is the atmospheric pressure at the altitude of the aircraft.

The airspeed indicator provides some very basic yet important information on the aircraft's relative velocity through the surrounding air. It helps the pilot establish optimum performance during takeoff and landing; it also shows an impending stall in most aircraft, though the less common angle-of-attack indicator gives a better warning of a stalling condition. Engine and airframe manufacturers use airspeed to establish structural limitations, such as airframe never-exceed speeds, flap and gear operating speeds, and propeller harmonic vibration speeds and rpm's.

Airspeed information is also used for operational calculations, such as determining groundspeed, and it is the controlling factor in optimum climb/descent efficiency for given conditions such as best rate, best angle, minimum controllable airspeed, and glide.

Fig. 29-1. *Maximum allowable airspeed indicator.* Photo courtesy of Kollsman Avionics Division.

In emergency situations, it can be an indicator of airplane attitude (power and prop being constant), increasing airspeed indicates descent, decreasing airspeed indicates a climb). Airspeed is a prime factor in establishing long-range cruise control, including fuel consumption, maximum range, and minimum time en route.

The construction of the airspeed indicator is simple (FIG. 29-2). The instrument case has a static air pressure vent but is otherwise airtight. A sealed diaphragm is inside the case; the bottom of it is permanently attached to the case, while the top is free to move. The diaphragm receives ram-air pressure via the pitot tube and expands or contracts depending upon the static pressure and ram pressure differential. The top of the diaphragm is connected by linkage to the indicator needle. The face of the instrument is calibrated in either knots (now standard for aviation) or miles per hour (often used in advertising to make the airplane speed ''look better''). (One knot means one nautical mile per hour; it is redundant to say ''knots per hour.'' There are 6080.27 feet per nautical mile versus 5280 feet per statute mile; one knot equals 1.151553 statute miles).

Not much goes wrong with an airspeed indicator except physical damage to the system and icing. If the pitot tube (FIG. 29-3) is blocked by ice, bugs, or trapped water, the airspeed indicator functions like an altimeter. The ram air, trapped within the diaphragm, becomes a fixed, static pressure. Static air continues to flow into the instrument case, but its density would vary with the aircraft altitude. As the aircraft climbs, density decreases, causing the static air trapped in the pitot tube to expand the diaphragm, and thereby causes indications of increasing airspeed. Similarly, a descent would cause the density of the air blocked in the pitot tube to increase, leading to an indication of apparent decrease in airspeed.

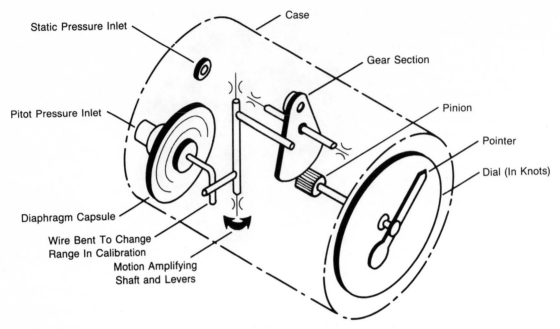

Static Pressure Inlet

Case

Gear Section

Pinion

Pointer

Dial (In Knots)

Pitot Pressure Inlet

Diaphragm Capsule

Wire Bent To Change
Range In Calibration

Motion Amplifying
Shaft and Levers

Fig. 29-2. *Simple airspeed indicator.*

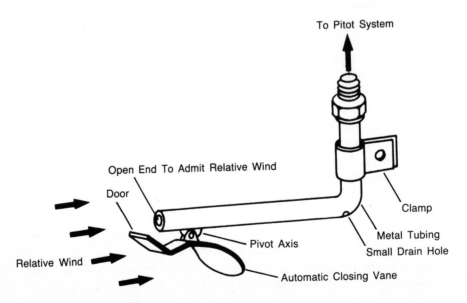

To Pitot System

Open End To Admit Relative Wind

Door

Clamp

Pivot Axis

Metal Tubing

Small Drain Hole

Relative Wind

Automatic Closing Vane

Fig. 29-3. *Pitot tube for a lightplane.*

Obviously, a blocked pitot tube is a dangerous situation. Picture a climb-out in instrument conditions. Apparently your airspeed is a little fast, so you instinctively ease the nose up just a bit, without even watching for the results. You continue climbing and again notice your airspeed is a little bit fast, so you ease the nose up once more. This time you get an unexpected stall. It is important to understand that when the pitot tube is blocked, the airspeed indicator no longer measures airspeed.

If the static port is blocked but the pitot tube remains clear, the airspeed indicator becomes an altimeter in reverse, showing an airspeed increase with decreasing altitude. When in areas of potential icing, airspeed corrections should be observed whenever a change of pitch or power is implemented. If the problem is the result of pitot-static blockage, the error will continue to increase without any sign of the correction having been made. Similarly, always maintain a complete instrument scan; all instruments should agree there is an airspeed problem; otherwise there is an indicator problem.

The airspeed indicator is designed to be most accurate at sea level on a standard day (59 degrees Fahrenheit); indicated airspeed is at best fraught with errors. It is subject to various system errors, leaks and errors resulting from improper pitot tube and static source placement. To compensate, the airframe manufacturers give us calibrated airspeed (CAS) which is corrected for system errors. Appropriate airspeed corrections are presented in the pilot's operating handbook (POH) for different conditions of flight, the most common of which are with gear and/or flaps up or down.

Because you seldom operate at sea level and occasionally fly higher-speed aircraft, you must also be concerned about compressibility effects on CAS. As density decreases, compressibility problems increase; but it isn't until approximately 10,000 feet that aircraft flying at 250 knots or faster begin to have serious airspeed indicator problems. At that point, it is necessary to compute equivalent airspeed (EAS), which is CAS corrected for density.

High speed, high altitude airplanes encounter such high ram air pressures that the "ram effect" causes the air to compress, affecting the airspeed indication. This is a well-known phenomenon that affects all aircraft in the same way. Charts provide compressibility factors based on altitude and CAS. Incidentally, this compressibility error applies only to the standard airspeed indicator; true airspeed indicators and Machmeters automatically correct for the lack of air compression.

All aircraft are affected by decreased atmospheric density or what pilots call "high density altitude." The less dense the air (whether the result of high altitude or hot temperature), the greater the error between CAS and true airspeed (TAS).

A calculation must be performed—taking temperature and pressure into consideration—to convert CAS (or equivalent airspeed, if applicable) to TAS. Some pilots buy true airspeed indicators that do the conversion automatically; others buy a less expensive basic airspeed indicator with a rotating dial around the outside. The dial, when manually set to the correct pressure altitude and outside air temperature, gives the pilot the TAS conversion. This falls into the same category as the manually rotatable ADF card; it's not a radio magnetic indicator (RMI), but it sure beats a fixed-card ADF! Those who can be happy with a reasonable TAS estimate should add 2 percent of indicated airspeed for every 1000 feet of altitude. When you take TAS and correct it for the ambient wind, you come up with the actual speed of the aircraft over the ground (groundspeed).

SENSITIVE ALTIMETER

To the uninitiated, the altimeter can be a very misleading instrument (FIG. 29-4). All it really does is measures the weight of air above the aircraft; it has no way of knowing how much air is below it. If it did measure height above ground, imagine flying at a constant altitude; the indicator needle would bounce up and down every time it passed over a hill, rock or building. Instead, you fly at a constant pressure level using an altimeter that is adjusted manually before takeoff to closely approximate sea level.

Airport elevation is, in fact, listed as feet above sea level. The trick is for the pilot to get an accurate altimeter setting from someone on the ground at the destination airport. If someone at the airport issues the correct information from an accurate instrument, the pilot puts in the correct altimeter setting. If there are no fast-moving cold fronts in the area to play havoc with the pressure, the altimeter should show airport elevation as the airplane touches down.

Older altimeters had one diaphragm (*aneroid*) and one hand. Their entire range was perhaps only two revolutions of the instrument and they were not very accurate (FIG. 29-5). Modern altimeters, known as "sensitive altimeters," have two or three aneroids (FIG. 29-6). If you hold an altimeter over your head and read it, then put it on the floor and read it again, you will notice, if you're of average height, the longest hand indicates a change of approximately 5 feet.

Of the three major types of altimeters—three-pointer, drum pointer and counter pointer—the most common in light, general aviation aircraft is the three-pointer (FIGS.

Fig. 29-4. *Pneumatic encoding altimeter.* Photo courtesy of Kollsman Avionics Division.

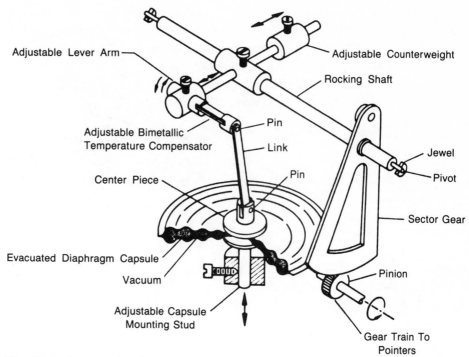

Fig. 29-5. *Basic altimeter mechanism.*

29-7, 29-8 and 29-9). With this instrument, the longest hand registers 1000 feet in one revolution, with each number around the dial equaling 100 feet. The wider but shorter hand registers 10,000 feet in one revolution and each number signifies 1000 feet. The smallest hand would register 100,000 feet if it ever made a complete revolution; each number it points to is read ×10,000.

The one thing all altimeters seem to have in common is the ability to be misread. Numerous accidents have resulted from altimeter misreadings, particularly the 10,000-foot indicator. More than one airplane has been cleared to one altitude, such as 12,000 feet, and ended up at another—2000 instead!

Because the altimeter must be compensated for non-standard atmospheric pressure if it is to indicate the aircraft's true altitude, there must be some way to adjust it (temperature variation is automatically compensated for with an internal bimetallic strip). The pilot dials in the local aviation altimeter setting, which simultaneously adjusts the drive mechanism to set the altimeter to compensate for non-standard (sea-level) conditions. An entire generation of pilots have called this the Kollsman Window without knowing why—it was the Kollsman Instrument Company who invented the process. In fact, Kollsman is the unofficial grand-daddy of altimeters, having invented the first truly reliable altimeter in 1928. According to Vernon Maine of Kollsman Instrument Company, all altimeter

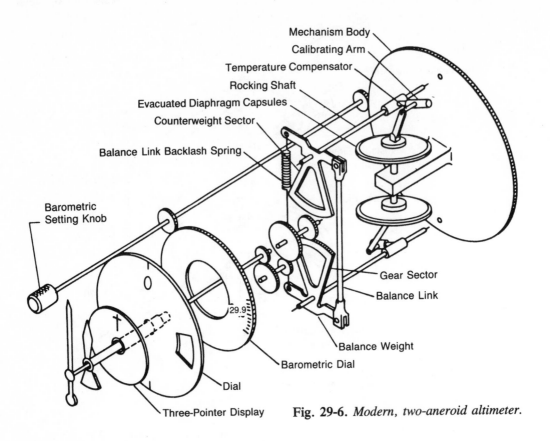

Mechanism Body
Calibrating Arm
Temperature Compensator
Rocking Shaft
Evacuated Diaphragm Capsules
Counterweight Sector
Balance Link Backlash Spring
Barometric Setting Knob

Gear Sector
Balance Link
Balance Weight
Barometric Dial
Dial
Three-Pointer Display

Fig. 29-6. *Modern, two-aneroid altimeter.*

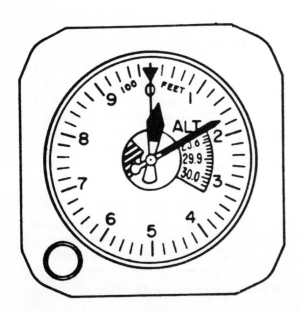

Fig. 29-7. *Three-pointer altimeter.*

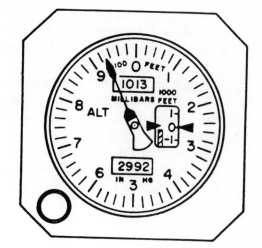

Fig. 29-8. *Drum-pointer altimeter.*

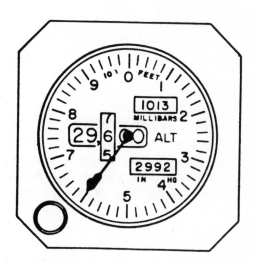

Fig. 29-9. *Counter-pointer altimeter.*

manufacturers now have barometric pressure adjustments. Regardless, Kollsman still is a leader in the industry with a reputation as long as its history.

With the aircraft sitting at sea level in standard conditions (59 degrees Fahrenheit, 29.92 inches Mercury (Hg)), the altimeter should read 0 feet if it is set at 29.92 inches Hg. For example, if while sitting on the ramp, the pressure drops to 29.42 and the pilot does not change the altimeter setting, the altimeter will climb to 500 feet—when the actual altitude is 0 feet (29.92 − 29.42 = .50) A change of

.01 inch Hg   =   10 feet

.10 inch Hg   =   100 feet

1.00 inch Hg   =   1,000 feet

Hence, the indicated altitude is 500 feet greater than the actual altitude. Therefore, the pilot should always keep the altimeter set to a current source within 100 nautical miles of the aircraft's present position and always update the altimeter for each point of intended landing.

The altimeter setting, though often called barometric pressure, is not the same thing that is available from a local weather station's barometer. Despite the same scale (inch Hg) and occasionally similar readings, they are in fact different and only an aviation altimeter setting should be used.

There are four basic types of altimeter errors: scale, friction, mechanical, and hysteresis. *Scale error* is the result of the aneroids not responding uniformly to the local pressure difference. This type of error is irregular throughout the instrument's range and difficult to predict. A greater margin for altimeter error must be taken into account during high-altitude operations.

Friction error is inherent whenever there are moving parts. Most commonly, the 100-foot pointer will "hang up." This tends to be less of a problem in reciprocating-engine aircraft than in jets, because they vibrate sufficiently to keep the pointer moving easily. Jets, which have smoother running engines and minimal vibration, use instrument panel vibrators to help overcome friction error.

When flying straight and level, if you are doing an exceptional job of holding altitude, it is a good rule of thumb to gently tap the instrument—you may be in for a surprise! If you know the altimeter has 'hung up' but a gentle tap doesn't release the pointer, simply twist the altimeter setting knob. Hard-to-twist knobs may indicate internal instrument corrosion, probably the result of static source moisture. If the pointer still doesn't move, static-source blockage, especially icing, would be the next logical suspect.

Mechanical error is the result of misalignment of gears or slippage of gears or linkage. For the most part, if the altimeter gives accurate indications on the ground during preflight, it will probably be mechanically sound during flight. Actual inflight, mechanical failure is rare, provided the pilot follows up on problems found during preflight. Occasionally, the set screw in one of the altimeter's hands works loose and the needle drops straight down. If it is the long needle that fails, one revolution is 1000 feet, so the medium needle will provide a reasonable indication of altitude within 200 feet. Understandably, it is not a good idea to be flying low approaches in this situation. If the medium length, fat needle works loose, you only need to keep track of the revolutions of the long needle to assure knowledge of altitude in thousands of feet. In case of a complete altimeter failure, there are two other last resorts. If you have a constant-speed prop, you can set the friction lock on the throttle and prop so they won't accidentally move and use the manifold pressure gauge as a crude altimeter. In a pressurized aircraft, you can depressurize the cabin and use the cabin pressure indicator for a fairly effective altimeter.

Hysteresis is a result of the elastic quality of the aneroids. After maintaining a constant altitude for a long time, the aneroids require a little time to respond to quick changes in altitude. While typically this was not a very significant problem, it has been all but eliminated in modern altimeters.

Absolute altitude is the altitude above the terrain directly below the aircraft. *Pressure altitude* is for reference purposes and is the altitude above a standard datum plane (sea level). *Density altitude* is pressure altitude corrected for temperature and is the same as pressure altitude only when ambient conditions are standard. *Indicated altitude* is whatever is displayed on the altimeter, while *calibrated altitude* is indicated altitude corrected for installation error. Then, we have *true altitude*, which is calibrated altitude corrected for non-standard atmospheric conditions; theoretically, it is your actual height above mean sea level.

Finally, there are *flight levels*, where the altitude of constant atmospheric pressure relates to the standard datum plane. All aircraft at or above 18,000 feet fly flight levels by adjusting their altimeters to 29.92 inch Hg. If all of this sounds confusing, from a practical standpoint it isn't. When flying below 18,000 feet, use the current altimeter setting from a reliable source within 100 miles of your location; above 18,000 feet, set in 29.92 inch Hg. If you are going to fly IFR, both the altimeter and static system must have been inspected within the preceding 24 months.

VERTICAL VELOCITY INDICATOR

The purpose of the vertical velocity indicator (VVI), sometimes referred to as the vertical speed indicator, is to indicate the aircraft's rate of climb or descent. Measuring the rate of change in ambient static pressure, the instrument (FIG. 29-10) is calibrated in positive and negative feet per minute. It also serves as a reference for level flight.

The VVI is a differential pressure instrument similar to the airspeed indicator. It has an aneroid, vented to static pressure, inside the instrument case and it operates a needle

Fig. 29-10. *Vertical speed indicator.* Photo courtesy of Kollsman Avionics Division.

through a series of gears and levers (FIG. 29-11). Instead of ram-air pressure, the case is vented to static air that passes through a calibrated restrictor. As the altitude changes, the diaphragm responds immediately to the free-flowing static pressure. The case pressure, through the restrictor, has an inherent airflow lag that artificially creates a pressure differential. Manufacturers have developed mechanical compensations for the retarding effect of cold temperatures on the airflow through the restrictor and to the inherent lag while differential pressure builds. Otherwise, the VVI has changed little over the years.

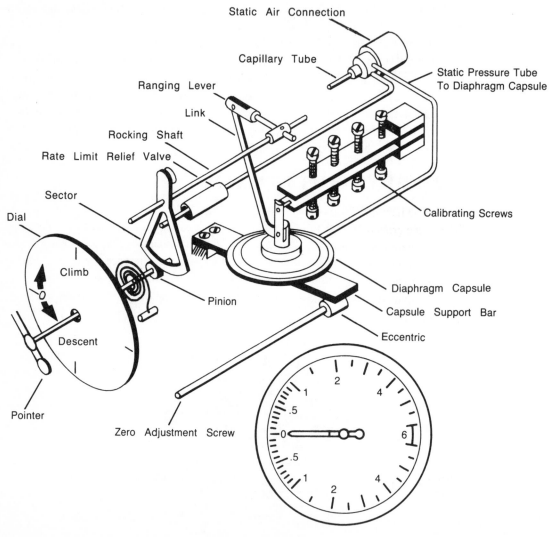

Fig. 29-11. *Schematic of a vertical velocity indicator.*

Used as a supporting trend instrument, the VVI helps the pilot establish constant rates of climb and descent and provides additional information regarding level flight.

PITOT TUBE

Named for the French engineer Henri Pitot (sounds like "pea-toe"), who invented it, the pitot tube measures the flow of fluid. In the airplane, it measures dynamic (ram) air pressure. According to Eric J. Daiber, president of Aero Instrument Co., there is less total pressure error if the pitot tube is *not* located in a wake, boundary layer, or region of supersonic flow. An extended boom in front of the fuselage probably is the most effective area, but it can complicate ground operations.

Typically, the pitot tube on general aviation aircraft is under the wing or, for larger aircraft, on the nose area. Often a small drain hole is added to permit rain to pass through; sometimes a sheet-metal vane is added to keep insects out of the tube when the aircraft is on the ground (airflow swings it open in flight; on the ground, gravity closes it).

Pitot tubes range in type from a simple bent piece of tubing that extends into the relative wind, to a pitot head with static source and heat (FIG. 29-12). Pitot anti-icing is accomplished by electrically induced heat around the tube, a method developed in the 1940s by Aero Instruments for the U.S. Navy.

The most common cause of pitot failure is actual physical damage to the pitot tube itself. The next most common cause is heater burn-out, primarily the result of using pitot heat on the ground. Third would be burning a pitot cover onto the tube, which is a particular problem with pitot-static probes because the plastic can literally melt into the static port.

Functionally, ram air enters the pitot tube and is brought to a complete stop, allowing pressure to build up to total free-stream pressure or "head" pressure. It then is transmitted to the diaphragm inside the airspeed indicator. Head pressure equals dynamic pressure plus static pressure. Therefore head pressure minus static pressure equals dynamic pressure or indicated airspeed.

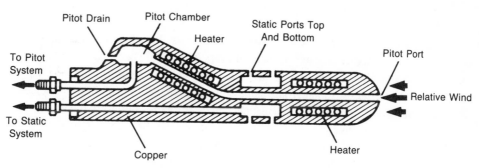

Fig. 29-12. *Heated pitot-static tube.*

STATIC PORT

The aircraft static port senses the atmospheric pressure at the aircraft's altitude. Ideally the port is located in the boundary layer of the fuselage. The column of air in the static port should be perpendicular to the local air velocity, the latter of which is exactly equal to the free-stream velocity of the airplane. If it is exposed to rushing air—from turbulence, for instance—the resultant ram-air pressure will cause the static pressure to increase and/or vary. Airflow passing by that has a greater velocity than free-stream air causes lower-than-static pressure. To compensate for problems caused when side-slipping the airplane, two or more static ports are located on opposite sides of the airplane and are cross-coupled to create an "average" static pressure.

Sometimes the static port is co-located with the pitot head. Don Reynolds of Rosemount, Inc., a manufacturer of pitot-static systems, explained the difference between the two types of static-source mounts:

"In the fuselage-mounted static source, accuracy decreases as airframe longevity increases, because any dings, dents, etc., around the static source changes the boundary layer flow and affects airspeed and altitude. Static is the most common source of problems, not pitot. Pitot-mounted static sources are located on a finely machined surface on the pitot probe itself and the airframe condition hardly affects it. This type of system remains accurate significantly longer than fuselage-mounted static systems."

It is most common that static ports located on the side of a fuselage will produce errors that vary with the configuration of gear and flaps, angle of attack, airspeed, and even the type of aircraft. For the light, general aviation aircraft however, the cost of a finely machined pitot-static head could outweigh the benefits. Consequently, most of them have the traditional flush-mounted, fuselage static system.

ALTERNATE STATIC SOURCE

An alternate static source on the aircraft is a must for the IFR pilot. Static-system failure and icing are not a problem when you can reach down under the instrument panel and twist a small knob that activates an alternate static source located where icing will not form.

Many pilots (even aircraft owners) haven't the faintest idea where the alternate source is located or even whether there is one in the aircraft. This should be a mandatory checkout for all instrument pilots. They also should test the system in flight at least monthly. In an unpressurized aircraft, the alternate static source is typically located inside the cabin where the ambient pressure is less than that outside the aircraft—therefore, the instruments will have errors. The operating handbook should detail the difference. If not, the pilot should become familiar with the errors in a controlled training situation.

If your aircraft has no alternate static system and the primary fails, the attitude indicator becomes a primary instrument and should be used with a power setting known to give appropriate airspeed for the desired phase of flight.

It also is possible to break the glass of the VVI to let static air into the system(airspeed and altimeter will also work, but they are more important instruments and permanent damage could occur when you break the glass). If you are fortunate enough to have copilot instruments, then break one of them if they are on the same system. If they are on a different system, you might consider sliding over to the right seat and finishing your flight from there! One note of caution about breaking the instrument in a pressurized aircraft: you must depressurize the cabin, or the instruments will read incorrectly because of high cabin pressure.

PREFLIGHT

The walkaround inspection should assure removal of the pitot tube cover if you use one. Also check the pitot tube, drain, and static ports for blockage, general condition, and alignment. Especially check the static ports for tape or other protection after the aircraft has been washed or painted. For flight into instrument conditions, turn on the pitot heat and feel the tube with your hand (be brief, both to save the heater element and your skin— they heat up fairly quickly).

The cabin check should include the current altimeter setting that reflects the airport elevation correctly; an error in excess of 75 feet should be cause for grounding. Both the airspeed indicator and VVI should read 0 prior to engine start. During normal taxi, all three instruments should remain fairly constant, aside from a little jiggling of the needles. If there are noticeable changes in airspeed, altitude, or vertical speed during taxi, the instruments are in trouble.

During takeoff, the airspeed indicator should come alive fairly early in the run. If the aircraft has two airspeed indicators, cross-check them when yours reaches the bottom of the white arc (a fairly easy-to-recognize position when you glance across the instrument panel to the other one.) If they are not in agreement, something is wrong. The vertical speed should reflect a climb promptly after leaving the runway, but there always will be a slight inherent delay. The altimeter should show upward movement within a couple of feet.

Questions of accuracy regarding pitot-static instruments are best resolved on the ground, so the sooner you admit there might be a problem, the earlier you can make the decision to abort a takeoff or avoid flying into IFR conditions.

30

Will Your ELT Work
When You Need it?

At best, search-and-rescue (SAR) is a hit or miss proposition. It wasn't all that long ago when the search sometimes claimed more lives than the downed aircraft. Rescue was often delayed by poor reporting procedures, and the actual search often had airplanes flying haphazardly over suspected areas, sometimes into each other!

The FAA sought to shorten the time between accident and rescue because many victims survived the crash only to die of exposure and hunger. At one point, it was determined that 50 percent of the persons who were successfully rescued were found in the first 12 hours; 25 percent were found within the second 12 hours. The odds weren't good if you had to wait longer than that. According to statistics compiled by the Air Force Rescue Coordination Center, about 35 out of 100 occupants of downed aircraft survive the impact. Twenty-one of the 35 are injured in some manner, but after 24 hours only 7 of the 35 will be alive!

On December 29, 1970, an amendment to the Federal Aviation Act of 1958 was voted into effect. It required most U.S. civil aircraft to have an FAA-approved emergency locator transmitter (ELT) installed and operating. The exceptions, as they pertain to general aviation, are listed in Federal Aviation Regulation 91.52, and are as follows:

1. Turbojet powered aircraft.

2. Aircraft while engaged in scheduled flights by scheduled air carriers . . .

3. Aircraft while engaged in training operations conducted entirely within a 50-mile radius of the airport from which such local flight operations began.

266

4. New aircraft while engaged in flight operations incident to the manufacture, preparation, and delivery.

5. Aircraft while engaged in flight operations incident to design and testing.

6. Aircraft while engaged in flight operations incident to the aerial application of chemicals and other substances for agricultural purposes.

7. Aircraft certificated by the Administrator for research and development purposes.

8. Aircraft while used for showing compliance with regulations, crew training, exhibition, air racing, or market surveys.

9. Aircraft equipped to carry not more than one person.

10. An aircraft during any period for which the transmitter has been temporarily removed for inspection, repair, modification or replacement, subject to the following: No person may operate the aircraft unless the aircraft records contain an entry which includes the date of initial removal, the make, model, serial number and reason for removal of the transmitter, and a placard is located in view of the pilot to show ''ELT not installed''; and no person may operate the aircraft more than 90 days after the ELT is initially removed from the aircraft.

The subject of required ELTs was very controversial at the time of its inception, and in some respects it still is. Over 90 percent of the distress signals are erroneous. Eighty percent of them originate at airports, and once they are finally found, the searchers typically can't contact the aircraft's owner. Sometimes they can't even get into the hangar. One ELT was found, after an extensive search, in the pilot's flight bag located in his hotel room.

I asked Richard A. O'Neill, Vice President of Marketing for Dorne & Margolin, Inc., if he thought ELTs were a good thing. ''A lot of people are alive because of ELTs. I don't know if the ELT concept ever really achieved what the FAA wanted, but I suspect if you ask someone who was saved, they'd think it has.''

Despite the problems, the statistics do tell an important story. In aircraft equipped with an ELT, the average time to locate the aircraft in a real accident situation was 22.2 hours. There was a survivor rate of 37 percent and it took an average of 19 flight hours to locate them. In aircraft without ELTs, it took an average of 4 days and 18 hours to locate the downed aircraft! The survival rate was 30 percent, and it took 127.3 flight hours to locate the survivors!

O'Neill was also quick to point out ''you have to remember that ELTs have made another contribution to the SAR process. Before ELTs, there were a lot of accidents and deaths resulting from the search process itself. The ELT has significantly helped increase the safety factor for the SAR teams.''

Most ELTs are self-contained, portable, hand-held devices that mount in the aircraft. Weight ranges from 1 to 5 pounds with the battery pack installed. Some have remote antennas mounted on the fuselage, some have antennas on the unit itself, and some have both types. When the aircraft is subjected to a G load in excess of the permissible range, such as in a crash, the ELT is activated by a "g" switch that is essentially a vibration sensor.

According to Charles A. Koster, president of Pointer, Inc., there are two principal types of activation methods: *rolamite switch* and *sliding mass*. The rolamite switch is a "rolling mass" type of switch. Under the influence of high g's, the switch mechanically closes the activation switch and the ELT begins to broadcast its warble alert. This type of switch is used by Narco and Pointer.

The sliding mass system, such as that used by Dorne & Margolin, is a spring-loaded, instantaneous type switch. The switch closes the electrical contacts which then hold themselves closed once contact has been made. Originally, there were some problems with static electricity accidentally activating the system, but the bugs have long since been worked out and either system provides adequate results. In either case, once the ELT is activated, it transmits a distinct, "wow-wow" modulated tone signal on 121.5 and in most cases also 243.0 MHz. The catch is that someone has to be listening.

INSTALLATION CONSIDERATIONS

With more than 200,000 ELT units installed in the U.S. alone, manufacturers have learned a few things about installation. A study of accidents reveals some interesting details about survivability. The tail section remains relatively intact approximately 84 percent of the time, but the nose only survives 1 percent and the cockpit only 2 percent of the time! Of course the ELT has a much higher survival rate in any of those areas, but that gives you an idea about the best place to mount it.

There are other considerations, too, such as potential for fire damage. And don't forget not only must the ELT survive but so must the antenna and connecting cable if it is externally mounted. Therefore, it is recommended that the ELT have an attached and externally mounted antenna and the unit be located as far aft in the empennage as possible (or immediately forward of it). Incidentally, 33 percent of all crashed aircraft end up inverted, so be sure the system is capable of functioning while inverted!

COMMON ELT-RELATED PROBLEMS

There have been definite ELT problems over the years. Probably the most common problem is inadvertent activation. The "g" switch can be activated by excessive vibration such as a hard landing, bouncing the aircraft off the taxiway, slamming the baggage door or towing the aircraft too quickly over uneven surfaces. Newer units, under the revised Technical Standard Order, should fare better in this area as the "g" switches have higher tolerances. Incorrect location of an ELT can also lead to problems. Such common locations as baggage compartment and rear seat tend to make the unit more vulnerable to inadvertent activation. Bouncing bags or stretching arms have often caused ELTs to go off. Make sure the proposed location provides adequate protection but still permits easy access to the unit so it can be checked when necessary.

If accidental activation is a problem, so too is lack of activation at the appropriate time. Always use the proper batteries for replacement. Whenever replacing batteries with anything other than original equipment, contact the ELT manufacturer for approval. It really isn't a sales gimmick to talk you out of Brand X; tests have shown that many substitute batteries won't activate the ELT, especially when they get a little bit older. Don't let advertising or even a TSO number fool you—check with the original manufacturer. What happens is some batteries produce a passivation layer as they age (this is especially true for magnesium type cells). The layer can cause a delay of over 1 second before the cell achieves rated voltage. ELTs with electronic latching circuitry can be activated by a crash pulse of less than 0.1 second, but if battery voltage is too low, the unit simply won't activate. It is important to point out however that modern ELTs with such circuitry probably don't require maximum voltage to activate—but why take chances? The problem can be totally avoided by using alkaline (zinc-manganese dioxide) cell batteries or any battery the manufacturer approves. Incidentally, the reason you might want the magnesium battery is because they have a longer shelf life and provide sustained high power for a longer period of time, especially in extreme cold.

If it sounds like there is a lot to worry about when considering batteries, you're right. The simple truth is that batteries are the most common cause of ELT problems and you just can't take good enough care of them. For extended storage, batteries should be refrigerated. Cold slows normal battery deterioration, increasing their shelf life. There are some risks associated with refrigeration though. When transferring the battery from cold to warm air, there is a chance of getting condensation within the cells or even within the entire battery pack. This leads to corrosion, premature battery depletion and the formation of conductive paths between cells that cause shorting and battery failure. Some manufacturers hermetically seal the cells to prevent such a possibility, but it is a good rule of thumb to always bring the battery to room temperature as slowly as possible and completely wipe it dry before installation in the ELT unit.

According to FAR 91.52 (d)(1) and (2), batteries used in the emergency locator transmitters required by Federal Aviation Regulations must be replaced (or recharged, if the battery is rechargeable) (1) when the transmitter has been in use for more than one cumulative hour; or (2) when 50 percent of their useful life (or, for rechargeable batteries, 50 percent of their useful life of charge), as established by the transmitter manufacturer under TSO-C91, has expired.

When ELTs were first required, they were powered by lithium ($LiSO_2$) batteries. The decision was not an arbitrary one; lithium batteries provide very high cell voltage, very long shelf life, very high power-to-weight ratio with more watt-hours per pound than their magnesium or alkaline equivalents, and they are available for a moderate cost. Perhaps one of the most significant reasons though was lithium's very good low temperature characteristics.

A properly maintained ELT with lithium battery could be fully functional at -40 degrees Celsius (C), the lowest temperature at which an unsheltered person has a chance of surviving a 24-hour period. With all that going for lithium batteries, it's no wonder they became the industry standard for the new ELT, but they were Trojan Horses waiting to be taken inside the aircraft.

It didn't take very long to discover that outgassing SO_2 reacts with moisture and becomes a highly corrosive agent. The resulting damage was a real surprise to many pilots. Vapor pressure inside the cells significantly increases as temperature increases, so there was a surprising loss of SO_2. Then there was the problem of reverse current sensitivity, which is not uncommon in an ELT battery, which dramatically increases cell internal temperature. The pressure can build rapidly enough to cause an explosion. Even if all goes well and the lithium cell vents as the pressure builds up, large quantities of SO_2 are expelled, typically into the cockpit, and the gas happens to be highly toxic. And lithium metal burns intensely in free air at 180 degrees Celsius. But even if the poor pilot survived the life expectancy of the lithium battery and finally replaced it, another problem is lurking: a discarded lithium cell is still capable of producing quite a bit of cyanide gas! Hence, it surprised no one when in 1979 the FAA issued an Airworthiness Directive requiring the removal of lithium batteries from aircraft.

In a study of non-functioning ELTs conducted in Arizona during 1980, it was discovered that many did not activate at the crash pulse required under TSO-C91. That same study also revealed that 14 percent of the ELTs examined had outdated batteries, while 6 percent had batteries that were dead. The importance of good preflight and preventive maintenance cannot be overemphasized. You should routinely have the ELT and battery checked; a 100 hour inspection is a convenient time, but certainly check them at every annual. Periodically test the unit by listening to 121.5 (and 243 if you have a radio capable of receiving on that frequency) and activating the ELT manually. This is an FAA-sanctioned test provided you only conduct it during the first five minutes of any hour and limit it to a few modulations.

In general, it is a good idea to turn the battery off when the aircraft is not in use, but always remember to include BATTERY ON in your preflight. While you're turning it on during preflight, also check the battery expiration date. If it is not readily visible, write it down on the ELT or next to it where it can be easily checked.

SEARCH AND RESCUE

Traditionally, the biggest stumbling block in the entire ELT and SAR process has been the human one—someone has to be listening! If someone knows you are missing, then SAR operations are set up and search aircraft will be listening for your ELT. Search responsibility goes something like this. If a flight plan is filed, the destination FAA Flight Service Station (FSS) begins a telephone search 30 minutes after the aircraft is overdue and no communication has been received amending the expected time of arrival. The specialist requests a ramp check of the destination airport, departure point, and airports enroute. It is very significant that the enroute airports checked are those that lie along the *flight-*

270

planned route, because the specialist cannot read your mind if you change routes. A telephone call is also made to your home and office if such information is available. This process takes anywhere from a half hour to an hour to complete, and if the aircraft is still unaccounted for, the FSS specialist then notifies the USAF Rescue Coordination Center. Once the Air Force is notified, they look at the *planned* route of flight and contact the appropriate units of the Civil Air Patrol (CAP) if over land, or the Coast Guard if at sea or over the Great Lakes, to begin the search. At the very best, an hour will have elapsed— more likely several hours—since the accident. Once notified, the CAP must marshall its entirely volunteer force, prepare search aircraft, brief the mission pilots, and set up a plan of operation. As organized and experienced as the CAP is, another hour or more might go by before the first search aircraft takes off, and there are no night search operations. If a flight plan is not filed, it could be days or even weeks before anyone reports a missing aircraft or person. More than one pilot has departed on an extended vacation into the wilds only to crash a few miles from home. With no one expecting to hear from the pilot for a week or more, the only hope is someone will be monitoring 121.5 on their radio. It is also important to note that according to regulations, an ELT battery is only required to operate continuously for 48 hours!

Fortunately, now someone is always listening. The Search And Rescue Satellite (SARSAT) project is a joint venture between Canada's Department of Communications, France's Centre National d'Etudes Spatiales, the U.S.'s NASA, and the USSR's Ministry of Merchant Marine. SARSAT has multiple satellites in low, near-polar orbits that listen for distress signals full-time. In addition to aircraft emergency locator transmitters, SARSAT also listens for shipboard emergency position-indicating radio beacons (EPIRBs). The ELT or EPIRB broadcast signal is received by a satellite which relays it to a ground-based local user terminal (LUT). The location of the beacon is determined by measuring the Doppler shift (relative motion) between the precisely known position of the satellite in its orbit and the fixed beacon. In the U.S., the information is then relayed to the Mission Control Center, who in turn relays it to the USAF Rescue Coordination Center. With the ELTs currently in existence transmitting over 121.5 and 243, the accuracy is easily 20 kilometers, but future, more sophisticated systems should have no problem with accuracy to within 2 to 5 kilometers. There are currently three satellites in orbit, and Koster told me ''we have been very pleased with the results so far.'' He seems assured that the future of successful SAR operations lies in space.

Like anything else, common sense and good operating practices will tremendously help the downed pilot get rescued. First and foremost, prior to flight you should advise someone at your departure point what your intended destination is, probable alternates, and your route of flight, especially if it won't be direct. The best course of action is to simply file a flight plan and notify an FSS if you deviate from your planned route or if you expect to be more than 30 minutes late (15 minutes for turbojet aircraft). Always remember to close your flight plan immediately after landing. You might be surprised to learn that if the CAP finds you sitting home reading the paper, you can be held liable for the fuel, oil, and communications costs of your false-alarm SAR operation.

When flying, if you have no other use for a spare com radio, keep it tuned to 121.5 and never assume someone else has reported an ELT signal! If you hear one, report the following to the FAA as soon as possible: altitude of reporting aircraft and where and when the signal was first heard, heard the loudest, and either faded or was lost. Time, weather, and circumstances permitting, consider doing a little snooping on your own. You can use the build-and-fade method of locating an ELT (FIG. 30-1).

THE BUILD-AND-FADE METHOD OF LOCATING AN ELT

This method operates on the simple principle that as the airborne receiver approaches the transmitting ELT, the signal increases in intensity. Conversely, the signal fades as the receiver moves away from it. When a search plane first picks up a signal (already in transmission) the searcher will be flying toward the ELT at some angle less than 90 degrees from the current heading. The pilot continues on the same heading until the signal begins to fade out. The pilot notes the distance flown while the signal was audible, then reverses course and flies back on the reciprocal heading for half the distance. Now the

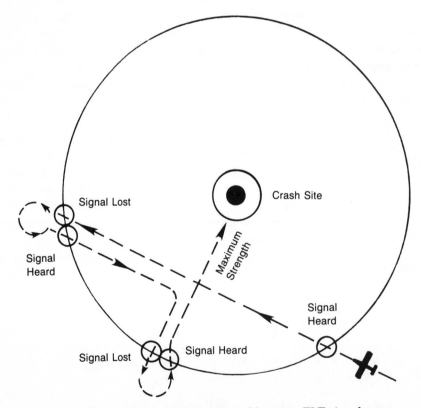

Fig. 30-1. *Build-and-fade method of locating ELT signals.*

pilot makes a 90-degree turn, either direction, and flies along this heading (or its reciprocal, as indicated by signal build or fade). When the aircraft reaches the point of greatest signal intensity, the pilot should be approximately over the transmitter. If directly over it, there will be an abrupt audible change and somewhat erratic signal. This position should be noted by reference to landmarks and coordinates on a sectional chart if possible. Whether or not the pilot spots evidence of aircraft wreckage, the FAA should be immediately notified to begin the SAR operation.

The accuracy of this signal-tracking procedure can be improved by using your squelch control. When an ELT signal is first heard, set the volume of the radio at your normal listening level and turn up the squelch to the point where the signal fades out. Then back off the squelch until the audio is just barely heard. If you track the signal with the squelch unchanged, it will reduce your required area search. On more sophisticated communications equipment, a "signal strength meter" indicates the position of maximum gain and reduces your search flying even more.

Be cautious if you are trying to help locate an ELT signal. It is easy to become mesmerized by events and more than one pilot has ended up adding to the problem. Not only have searching aircraft run out of fuel, but there have been cases where they have gotten bitten by the same meteorological conditions as the downed aircraft, such as severe mountain valley downdrafts. There is always the high danger of flying into terrain while your attention is diverted to directly below, and of course the potential for a midair collision is significantly higher than normal as other aircraft respond to the signal.

WHEN YOU'RE THE TARGET

If you find yourself as the object of a SAR effort, there are several things to keep in mind. After checking to make sure the ELT is alive and well, the first order of business is to take the necessary steps to ensure survival. If the conditions are hostile, protect yourself as necessary from the elements.

Remember that the ELT transmits omnidirectional, roughly in a circular pattern. They have been received as far as 100 miles away by aircraft operating at 10,000 feet. Transmission is line-of-sight and can be blocked if the antenna is under the fuselage or any other metal structure; even rough terrain can play havoc with the transmission signal. Of course a broken antenna is a serious problem, but there have been many cases where once an aircraft gets into the correct general area, it can home in on an ELT lacking an antenna.

If terrain appears to present a problem in signal broadcast, consider relocating the ELT, perhaps higher up in a tree or on a ridge. Finally, it is important to understand that a downed aircraft is extremely difficult to spot from both the air and ground. There have been many instances where searchers have passed within a quarter mile of a wreck and never seen it. Take whatever steps are necessary to ensure high visibility. Flashing with mirrors and glass are attention-getters, as is a smokey fire.

31

How to
Use and Maintain
the Magnetic Compass

While there have been many improvements over the years, the modern aircraft compass bears a striking resemblance to its earlier counterpart; certainly the fundamentals have not changed. Most science students have studied compass theory by floating a cork in a pan of water and placing a magnetized iron sliver or needle on top of it. Because the water has no static friction, the cork turns in response to the pull of the Earth's magnetic field on the north-seeking pole of the needle. Water does have sufficient friction however to prevent the needle from over-shooting. The cork also assures that the needle floats horizontally, a potentially significant problem the closer to the north or south pole the compass gets. This simple experiment, or variations of it, guided navigators at sea for centuries.

The real function of today's magnetic compass is one of redundancy. All but replaced by the directional gyro (DG), the compass has been relegated to the role of backup, occasionally being used to reset the DG to proper magnetic orientation. Unfortunately, compass knowledge has almost become obsolete too.

One of most neglected of all aircraft instruments, the compass provides the heading of the fore and aft axis of the aircraft relative to magnetic north. This should not be confused with either *course* or *track*. Course is a line drawn between two points while track is the actual movement of the aircraft with relationship to the ground.

Pilots hope their actual ground track will be the same as the course they plotted during preflight, but in reality the effect of wind usually makes that difficult. From a planning point of view, magnetic course plus or minus computed wind correction angle equal magnetic heading. But planning to use a compass and actually using it are two very different things. The effect of existing errors is so significant that a thorough understanding of all of them is necessary to make the compass a reliable piece of equipment.

EARTH AS A GIANT MAGNET

The Earth is a large magnet (FIG. 31-1). It has poles and a flux field, but it is a bit deceiving in that the true geographic north and south poles (those at the top and bottom of the globe) are not the magnetic poles. Magnetic north, located in Canada, changes position ever so slightly every year. That is one reason why maps and aeronautical charts are laid out according to unshakable true north.

The flux field around the Earth is nothing more than a giant version of a small magnet. Flux lines come vertically (an angle of 90 degrees) out of the south pole, bend around until they run parallel to the Earth's surface at the equator, then curve back and vertically reenter the Earth at the north pole. At points in between individual poles and the equator,

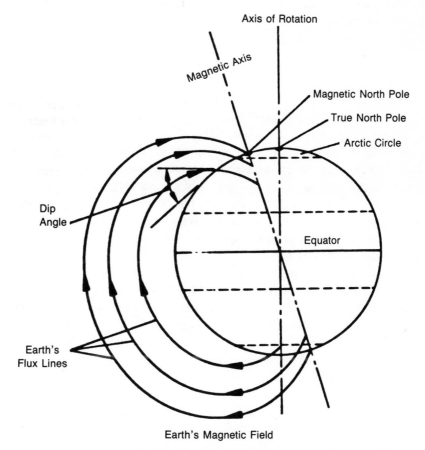

Fig. 31-1. *The Earth's magnetic field.*

flux emerges and enters the Earth at angles of less than 90 degrees. The imaginary angle between the flux and the Earth's horizontal plane is called the *dip angle*.

Imagine yourself standing thousands of feet up in the air with no magnetic objects within miles. Then suspend a long, rectangular magnet by a thread located right at the balance point. The first thing you would notice is that it aligns with the Earth's flux field, pointing toward the magnetic north pole. The next thing you would probably notice is that, with a single exception, one end of the magnet is pointing downward, as if out of balance. In the northern hemisphere, it would be the north-seeking end, which is incidentally the south pole of the magnet (remember, opposites attract). In the southern hemisphere, the south-seeking pole would be dipped. Only at the equator would the magnet appear to be balanced; parallel to the Earth's surface.

The greater the dip angle, the more severe the problem with compass accuracy. Taken to the extreme, as when directly over the north pole, the compass needle wants to point straight down into the ground. It would obviously be difficult to navigate with the needle pointing straight down, so knowing the direction of the lines of flux alone is insufficient. You need to know the direction of the flux lines relative to a horizontal plane. That defines the direction relative to magnetic north.

COMPASS COMPONENTS

Considering the simplicity of the concept, compass design is surprisingly involved. The modern compass is a compromise of numerous conflicting requirements. Though there is some variation in manufacturing, it is essentially built as follows (FIG. 31-2). The magnets, which you would expect to be big and strong, actually are not. A heavy magnet has a tendency to bob and move around excessively because of inertia. Also, a large magnet has too big of a flux field of its own that is greatly influenced by magnetic objects near it. By using two slender, long, parallel magnets, one on each side of the pivot, both the flux field and inertia are kept relatively small. It is the purpose of this very hard, pointed pivot to mount the magnets on as friction-free a point as possible so they can swing freely to align with the Earth's flux field.

The pivot fits into a virtually friction-free, very hard jewel cup. Here is yet another reason for moderation in flying technique. Hard landings, fast taxi over rough surfaces, and abrupt maneuvers in flight all cause bouncing and vibrations, the major cause of point blunting. As the point loses its sharpness, the compass loses its reliability. As a result, the compass card assembly is often mounted on a spring to minimize the problem, but the best preventive maintenance is still gentle handling of the aircraft. On larger compasses, the weight of the magnets and card assembly alone is sufficient to put a significant load on the pivot, so manufacturers often use a float with just enough buoyancy to relieve friction without actually floating the magnet and card assembly in the liquid.

The aviation compass card, typically a lightweight metal cylinder, has the cardinal headings depicted as N, S, E, W. The scale between headings is marked every 30 degrees

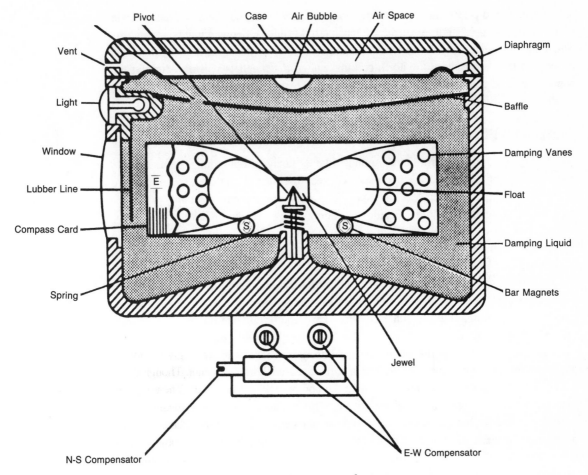

Fig. 31-2. *Aircraft compass elements.*

with the appropriate number. The last zero of these numbers is omitted so 30 degrees is marked by a 3; 180 by an 18. Between the numbers there are small tick marks indicating 5-degree increments.

Because the pilot reads the card from behind the compass, the markings are actually printed on the opposite side of the card. For example when flying north, the letter W is on the right hand side of the card, which is actually the east side of the airplane. Confusing, but logical. To aid the pilot in accurately reading the compass, a vertical wire or reference mark called a *lubber line* is placed on the front of the instrument behind a glass window. It corresponds to the longitudinal axis of the aircraft and the aircraft's magnetic heading falls directly under it.

To maintain the horizontal reference necessary to assure accuracy, the entire assembly (card, magnets, and float) must hang below the pivot point. If it sits too high, the magnets will succumb to dip error and the card will be excessively off the horizontal plane when flying in areas of high latitudes. On the other hand, if it is hung too low, it will be excessively pendulous, swinging easily during turns and acceleration. Even under the best of conditions, in the northern hemisphere the north-seeking end of the magnets will dip slightly; the opposite will be true in the southern hemisphere. To reduce the problems the compass is filled with a liquid. This is known as compass damping.

COMPASS FLUID

Filling a compass with liquid sounds like a fairly simple thing. Actually the requirements for this liquid are complex. To keep the size of the float to an absolute minimum, the liquid must have a high specific gravity. It must not get cloudy with age; very common for many liquids. Besides not being able to freeze or have a high vapor pressure, it must also maintain a relatively constant viscosity throughout extreme temperature changes to assure continuous damping of rocking and rotational oscillation of the card.

The fluid used in a compass must be both non-toxic and flame resistant. Originally alcohol was used, but eventually manufacturers went to both kerosene and trichlorethylene. Currently, it is common to use either acid-free kerosene or silicone fluids, because they not only have the dampening effect but also lubricate the pivot point.

In some compasses, to prevent overswing during magnet alignment (remember fluid has very low friction), vanes are used as a sort of sea anchor. But when rolling out of a turn, momentum keeps the fluid turning in the compass which tends to push the vanes (and the attached compass card) along with it. This phenomenon is very similar to the fluid in the middle ear which leads to vertigo under the same situation. To reduce this problem of momentum, holes are drilled in the vanes to allow the fluid to pass through while leaving sufficient vane to damp overswing.

Some small compasses do not have liquid dampening. These are damped by magnetic eddy current. To do this, the manufacturer uses the flux from the north-seeking end of the magnet. Because of the need for a higher-than-normal flux field, magnets with unusually high power-to-weight ratios are used. If doing away with the liquid sounds like a good idea, consider the consequences. The compass, which now has a greater flux field surrounding it, becomes highly susceptible to cockpit magnetic disturbances. Also the heavier weight and lack of lubrication lead to early pivot point dulling and decreased accuracy. The advantage of such a compass is that it is relatively inexpensive.

In the liquid-damped compass, the vane assembly is fitted carefully into a leak-proof case with a clear window. Then, the compass is filled with the liquid, making sure there are no air bubbles.

Because liquid volume varies with temperature and virtually all liquids produce some gas over time, a flexible, perforated upper baffle is installed. The holes allow air to penetrate into an upper chamber, keeping the compass card chamber filled with liquid. In fact, the upper chamber is commonly filled with air to provide a variable pressure compensator conceptually similar to a hydraulic system accumulator nitrogen precharge. Some manufacturers put a diaphragm at the back of the compass, which works fundamentally the same. Now that you understand how the compass works, it's time to consider operational compass errors.

COMPASS ERRORS

There are two types of compass errors: static and dynamic. *Static errors* are the result of deviation—compass accuracy degraded by local disturbances. The source of deviation can be anything in the cockpit that has a magnetic effect such as motors, any magnetic objects installed in the aircraft, any iron or steel structure, and actual magnets used in other equipment. There is also pilot-induced deviation such as laying a metal clipboard or a stopwatch on the instrument panel. And of course any current-carrying wire, especially one twisted in the shape of a loop can be influential; think of all those radio leads behind the instrument panel!

To minimize deviation, a compensator is located next to the north-seeking magnets. It consists of 2 sets of one or more small permanent magnets that can be adjusted to compensate for local disturbances. By turning the compensator screws, these small magnets are rotated and their flux fields interact predictably with the compass' magnets. One set is for north/south adjustment and a second for east/west. These should only be set by a mechanic, as the procedure is done for a specific compass and aircraft under specified conditions. The procedure is done entirely on the ground with the aircraft configured for normal flight, radios and electrical equipment on. Called "swinging the compass," it is accomplished in the following manner.

First the aircraft is taxied onto a compass rose. Many airports have them painted somewhere on a taxiway or ramp surface. The aircraft is faced toward magnetic north according to the compass rose. The mechanic adjusts the north/south compensator until the compass reads correctly. Then the aircraft is turned toward magnetic east and the east/west compensator is adjusted until the compass reads correctly. Similarly, the aircraft is turned toward magnetic south, but this time the compensator is adjusted until the compass reads halfway from its current reading to where the compass would actually read south, because you are now splitting the error difference between north (which was set accurately) and south. The aircraft is again turned, this time toward magnetic west and the difference is split between west and east. Going back to north again, the process of "splitting the difference" is repeated for north, south, east and west one more time. The result is the best compromise possible for the deviation that exists in the cockpit. Other methods of swinging a compass without a rose include using a hand-held direction finding instrument called a *pelorus* or a true, north-seeking gyro compass.

At best, compensators remove only part of the deviation, so after the compass has been swung it is important to document the error for the pilot. This is done by filling out a compass correction card. To do so, the aircraft is turned at 10-degree increments according to the compass rose. At each increment, the actual magnetic compass heading is read and the difference (deviation) is listed on the correction card. For instance, when the aircraft is positioned on the compass rose at an actual magnetic heading of 10 degrees, if the compass reads higher than 10 degrees, say 12, the difference is displayed on the compass card as STEER 12. This indicates to the pilot that to fly a magnetic heading of 10 degrees, it is necessary to steer the aircraft to a compass heading of 12 degrees.

From a practical point of view, a compass in good condition is accurate only in straight-and-level, unaccelerated flight. Any other time it is plagued by any of several dynamic errors. The most obvious of these is oscillation, the erratic movement of the compass card as a result of turbulence or rough handling. Whether induced by the environment or the pilot, when the airplane bounces and shakes, the compass behaves unreliably. When flying by instruments, smoothness is the all-important word. Except for those flying over the equator where the Earth's magnetic field is horizontal, some errors are, practically speaking, ever-present.

Turning error, which is dominant when flying either north or south, affects the compass because in a turn, the compass card banks with the aircraft. In the northern hemisphere on a northerly heading, turning to the east for instance, the card banks to the right. Remember the farther north the aircraft is, the further the compass magnet dips down at its north-seeking (front) end. When the centrifugal force of the bank is added to the dip, the card swings toward the west (remember west is on the right side of the card). As the turn proceeds toward the east, this error causes the compass heading to lag way behind the actual aircraft heading. If the pilot rolls the aircraft out of the turn when the compass reads east, the aircraft will be way past the target heading. Therefore, it is necessary to anticipate this and roll out early.

On the other hand, if the aircraft is on a southerly heading, the error produces a greater heading change than actually occurs (compass leads actual heading change) and the pilot must allow for the extra heading change before rolling out. For those of you with sympathies that lie south of the Mason-Dixon Line, there is a little crutch to remind you of these errors that will warm your hearts: The south leads, while the north lags.

Because of the inherent difficulties this error can produce while flying solely by reference to instruments, use the DG as your primary direction instrument. However, DGs have been known to fail, and the compass might be your sole heading instrument. In such a case it is imperative to be proficient at timed turns using a combination of a rate-of-turn indicator and the magnetic compass. As a rule of thumb, the pilot should allow for a rollout to the desired heading proportional to the latitude where the flight is taking place. For instance, when turning from south to north at approximately 30 degrees north latitude, let's say a left turn, start rollout 30 degrees prior to north *plus* another 5 degrees to allow time to roll the wings level. This means you should begin your rollout at a compass indication of 035 degrees. When turning from north to south, let's say a right turn, you must fly past south, in this case, by 30 degrees *minus* the 5-degree rollout lead, or 205 degrees. So the trick is to add or subtract the latitude and allow an additional 5 degrees to stop the turn. If all this makes you think perhaps it is best to try to plan all your trips so you only fly on headings of east and west, read on.

Acceleration/deceleration error occurs on headings of east and west. By now you are well aware that the north-seeking end of the compass is already dipped down slightly when the aircraft is in the northern hemisphere. When the aircraft accelerates on an east or west heading, the aft end of the compass card tilts upward, causing the card to rotate toward north (again remember that the north indication is actually on the south side of the card). Therefore, it appears that the aircraft is turning toward the north when in fact it is simply accelerating toward the east. For the opposite reason, deceleration causes the card to tilt in the opposite direction and rotate toward the south. The way to remember this error is through the acronym ANDS: accelerate north, decelerate south. The only compensation

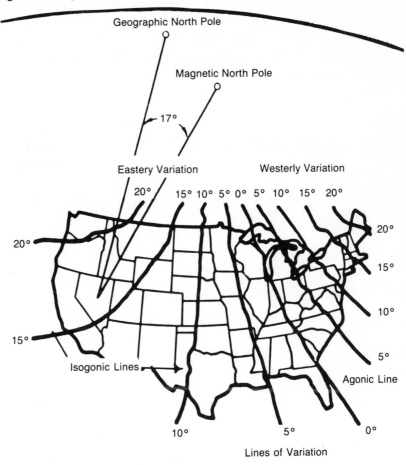

Fig. 31-3. *Lines of variation.*

is to assure you are in fact not changing your heading by closely monitoring the DG (if operative) or the turn and attitude indicators.

Variation, though not actually a compass error, is still something to be considered. Because magnetic and true north are not the same, a correction must be applied to the true heading to navigate by a magnetic compass (FIG. 31-3). This is because charts are laid out according to true north. One of the reasons for that is because the magnetic north pole shifts continuously. Even instrument charts are laid out according to true north, but the airways are depicted as magnetic because they emanate from the navaids that are laid out magnetically.

The angle difference between true and magnetic depicted on charts is called variation and the individual lines, called isogonic lines, are labeled either E or W (for east or west). The only exception to the rule is the one line that runs directly through both true and magnetic north. That 0 line of variation is called the *Agonic line*. Otherwise, you add W to the true heading to get magnetic heading and subtract E from the true heading to get the magnetic one. So, if you draw a line between two airports on your sectional chart and find that the true course is 310 degrees, you then look for the magnetic variation in that area. Let's say its 10E. 310 − 10 is 300 degrees. Therefore, you must fly a magnetic course of 300 degrees. But don't forget compass deviation. Now you must consult the compass correction card. Perhaps it says for 300 degrees, STEER 305. So you must actually fly a compass course of 305 degrees. All of this is assuming you are flying in the northern hemisphere. Turning and acceleration errors are exactly the opposite in the southern hemisphere.

INDEX

Edited by Lisa A. Doyle

Other Bestsellers From TAB

☐ **STANDARD AIRCRAFT HANDBOOK—4th Edition—Edited by Larry Reithmaier**

The classic guide to FAA approved metal airplane hardware has now been completely updated, revised, and expanded! First published as a general guide for aircraft workers, this extensively illustrated handbook has become an indispensable resource for A&P mechanics, technicians, students, pilots, factory personnel, and aviation enthusiasts. Anyone building, maintaining, overhauling, or repairing all-metal aircraft will find it an essential reference. Highlights techniques and procedures used in riveting, fastening, plumbing, and more. 240 pp., 239 illus.

Paper $8.50   **Hard $9.95**
Book No. 28512

☐ **THE ART OF INSTRUMENT FLYING—J.R. Williams**

This book addresses all elements of IFR flight using situations derived from the author's actual flight experiences. Each subject inherent to safe IFR operations is covered: instrument approaches and departures, weather, airspeed/altitude control, radio communications, charts, IFR flight planning, and spatial disorientation. 272 pp., 85 illus.

Paper $18.95   **Hard $24.95**
Book No. 2418

☐ **THE HELICOPTER—Keith Carey**

Veteran helicopter pilot Keith Carey has produced a fascinating book that provides a detailed description of the evolution of the helicopter from the first "semi-successful" versions to today's sophisticated machines. Meticulously researched and illustrated with more than 200 photographs and diagrams of how helicopters work, this is a reference that should not be missed by anyone fascinated with rotorcraft in particular, or aviation history in general. 224 pp., 227 illus.

Paper $12.95   **Hard $14.95**
Book No. 2410

☐ **WINGS OF THE WEIRD AND WONDERFUL—Captain Eric Brown**

The Guinness Book of Records lists Captain Eric "Winkle" Brown, the former Chief Naval Test Pilot and Commanding Officer of Great Britain's Aerodynamic Flight at the Royal Aircraft Establishment, as having flown more types of aircraft than any other pilot in the world! Though his test and naval flying writings are already internationally known, he has once more opened his flying logbooks to reveal some of the more unusual types of aircraft. 176 pp., 77 illus.

Paper $15.95   **Hard $19.95**
Book No. 2404

☐ **ABCs OF SAFE FLYING—2nd Edition—David Frazier**

Attitude, basics, and communication are the ABCs David Frazier talks about in this revised and updated second edition of a book that answers all the obvious questions, and reminds you of others that you might forget to ask. This new edition includes additional advanced flight maneuvers, and a clear explanation of the Federal Airspace System. 192 pp., 69 illus.

Paper $15.95   **Hard $19.95**
Book No. 2430

☐ **FLYING VFR IN MARGINAL WEATHER—2nd Edition—Paul Garrison, Revised by Norval Kennedy**

In this revised edition, you'll find technological information on such weather phenomena as wind shear . . . details on today's most advanced lightplane instrumentation including altimeters, airspeed indicators, vertical speed indicators, turn-and-bank indicators, and more . . . tips on the use of wing levelers and autopilots . . . and a practical look at the most advanced new technology in VHF navigation receivers, OBIs, and other navigation equipment including Loran C. 224 pp., 91 illus.

Paper $16.95   **Hard $21.95**
Book No. 2416

☐ **ARV FLIER'S HANDBOOK—Joe Christy**

Here's a practical, in-depth look at this increasingly popular new aviation category with a realistic assessment of the ARVs advantages and problems. It includes an overview of the ARV's available today—machines on both sides of the 254 lb./55kt. dividing line. Covered are costs, materials, construction standards, obtaining a license, insurance facts, even cost comparisons with traditional lightplanes. (A "fun" plane that will do 80 mph and costs about $12,000.) 192 pp., 76 illus.

Paper $10.50   **Hard $12.95**
Book No. 2407

☐ **I LEARNED ABOUT FLYING FROM THAT—Editors of *FLYING*®Magazine**

The editors of *FLYING* Magazine have selected the very best from their publication's most popular regular feature to come up with a series of sometimes humorous, always candid, flying stories. These exciting tales by dozens of high-time pilots—including airman-aerobat Paul Mantz, pilot-author Richard Bach, movie stunt pilots, and many others—all provide valuable flying lessons! 322 pp., 8 illus.

Paper $11.95   **Hard $13.95**
Book No. 2393

Other Bestsellers From TAB